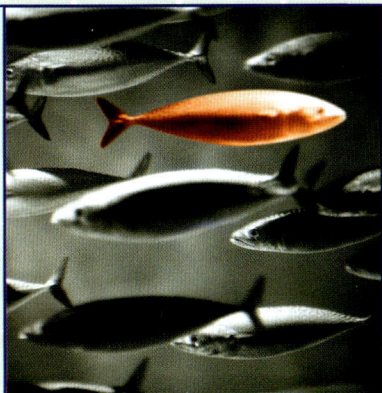

BE CONNECTED.

BE OPEN.

BE SUSTAINABLE.

BE ECONOMICAL.

BE YOURSELF.

BE WITHOUT WORRIES.

JUST BE.

英国版式
设计教程
（畅销版）

Design is a way of **looking** at and improving the **world** around us.

照片
剪切和尺寸调整

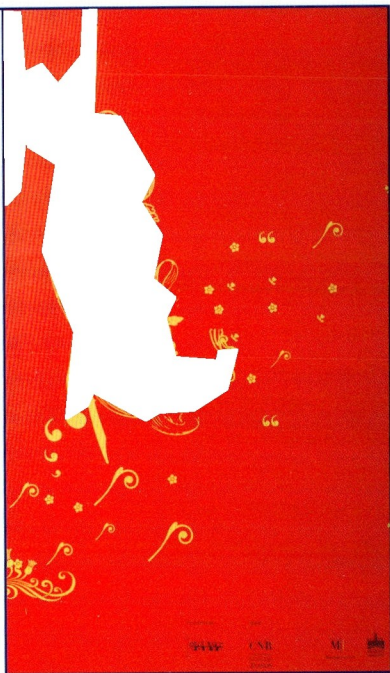

英国版式设计教程（畅销版）

[英] 戴维·达博纳 著

彭 燕 译

艾伦·斯旺《英国版式设计教程》的畅销版

上海人民美术出版社

图书在版编目（CIP）数据

英国版式设计教程：畅销版/（英）达博纳（Dabner, D.）著；
彭燕译.—上海；上海人民美术出版社，2014.1
　书名原文：How to understand and use design and layout
　ISBN 978-7-5322-8295-1

Ⅰ. ① 英… Ⅱ. ①达…②彭… Ⅲ. ① 版式-设计- 英国-
教材 Ⅳ. ①TS881

中国版本图书馆CIP数据核字（2013）第005874号

原版书名：How to understand and use design and layout
原作者名：David Dabner
©英国QUARTO出版公司

合同登记号：图字：09-2012-842号

英国版式设计教程（畅销版）
著　　者：[英] 戴维·达博纳
译　　者：彭　燕
责任编辑：钱欣明
技术编辑：季　卫
出版发行：上海人民美术出版社
　　　　　（地址：上海长乐路672弄33号　邮编：200040）
网　　址：www.shrmms.com
印　　刷：利丰雅高印刷（深圳）有限公司
开　　本：787×1092　1/12
印　　张：10.67
版　　次：2014年1月第1版
印　　次：2014年1月第1次
印　　数：0001-4000
书　　号：ISBN 978-7-5322-8295-1
定　　价：48.00元

目录

导言

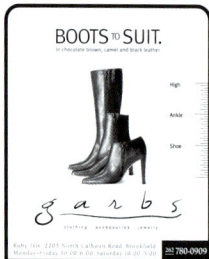

本书旨在讲述版式设计的全过程，既有助于初学者了解优秀设计所包含的要素，又能帮助经验丰富的设计师更新理念或进行艺术形式的新尝试。

本书分为两个部分：第一部分是设计的基本要素，包括设计师必须作出的诸如字体、色彩、构图、注释、照片等等的选择。第二部分是设计项目及类别，探讨了各种不同类型的设计，如标识、杂志、小册子、广告、网页设计等，并且对每一类型的设计都专门进行阐释。第一部分和第二部分中的每一个主题都包含了一段简短的讲解，并配以专业设计水准的优秀图例和供读者尝试的练习题。

全书的一条主旨是指引读者去解析自己的作品，能够客观公正地评价你所做的努力是学习过程中的重要部分：它不仅可以帮你成为一个好的设计师，还可以让你更清楚地意识到在任何一个设计项目中需要提出或回答的问题。同样，你也必须学会接受对你作品的批评，并且要积极面对。如果你固执己见，不考虑其他人的意见，那你可能就会止步不前。但是，对一项设计的任何评论都应有助于设计的改进。诸如"我不喜欢这个"之类的否定意见，没有提供任何建设性建议，反而会阻碍设计的进展，甚至能使你不知所措。

假设你已具有计算机操作技能且具备了进行练习的必需设施，你还要有一套常用的软件，才能达到本书的目的。

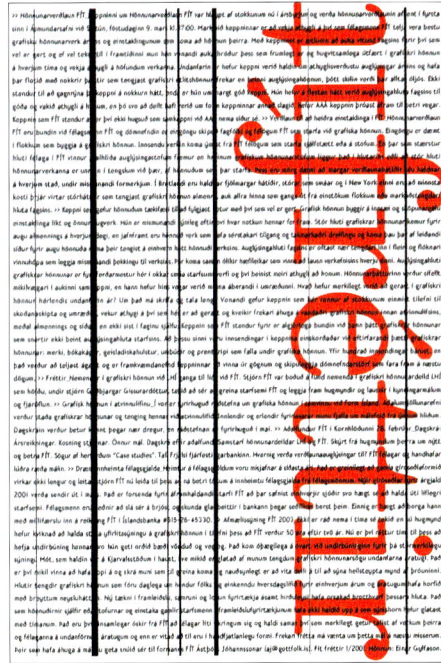

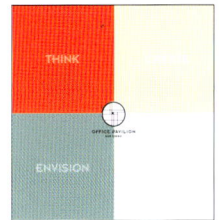

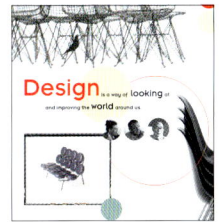

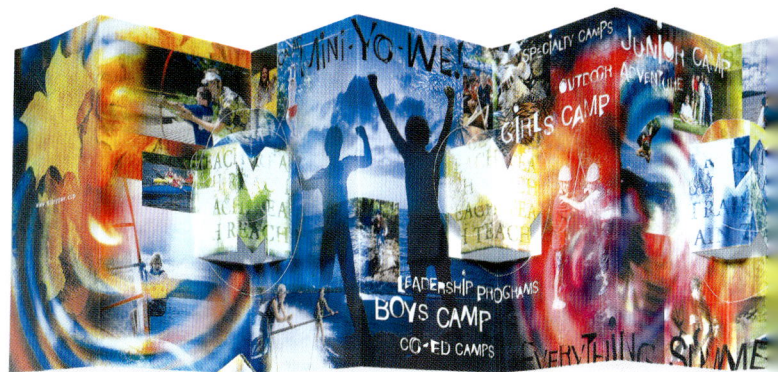

如何使用本书

第一部分

前面的几个自然段从理论的角度概述主题。在这个例子中的主题是"编排字体"。

在概述后面，你会看到"设计效果"部分，其中的内容是对下两页中略图的注释。

这些特别制作的示范作品对前面讲述的设计理念进行解释。每张图都标有编号，与"设计效果"部分中相应的说明语句对应。

每一部分都附有练习题以供读者练习前面提到的设计方法。

每道练习题均附有配图的范例答案。

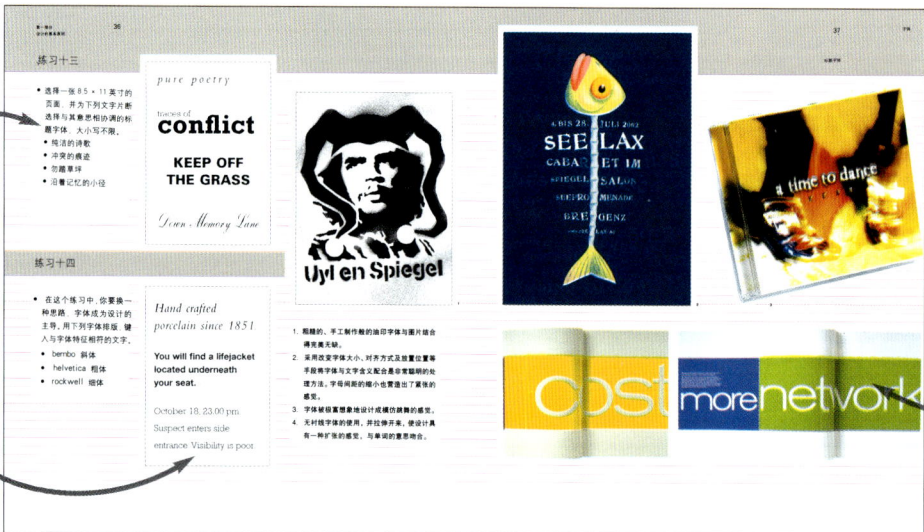

专业的范例一定会给你灵感。看看来自世界各地设计师们是如何设计出这些针对各种类别的精美作品的。

第二部分

每一类别设计项目的介绍中包括常见的实用操作技巧和使作品趋向专业化的可行建议。

在创造性设计练习中一试身手吧。练习中会提出设计的要求，接下来就看你的了。设计范例只是告诉你如何着手，而你将面临的挑战是做出自己的设计。不要忘了把你的作品给朋友和同事们看看，他们的意见和建议会让你进步得更快。

对每一项专业的范例进行深入地剖析，把设计中包含的各个元素分解开来，并对每个元素进行审视，将有助于理解设计师是如何将这些要素组合起来的，他为什么要这样组合。

来自世界各地的设计范例展示了专业设计师们对各类设计项目进行的尝试。

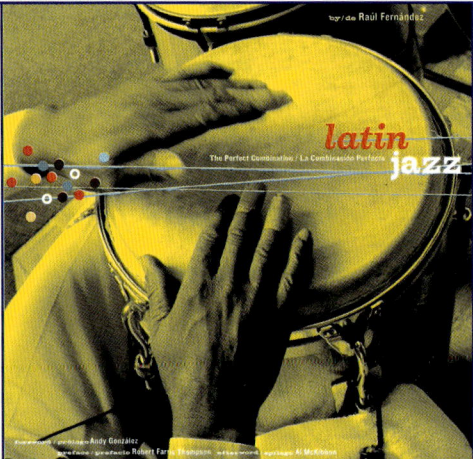

第一部分

设计的基本原则

在这部分中，图文并茂地阐述了设计过程中的各种元素，也即是最为基础的各种组成部分，如选图、选字体，再把它们组合在一起等。假如你肯花些时间，仔细阅读书中的每一页的话，对它们的阐述将会使设计对你而言易如反掌。

书中对设计中每一个组成部分都有一个简短的概述和详细的图例。后面还附有练习题及专业设计师对该题所做的范例解答，向读者展示他们是如何处理设计的基础问题的。通过揣摩这些范例，你可以看到实际设计过程中的取舍。

照片
剪切和尺寸调整

设计中首先要考虑的问题是设计作品的形状和尺寸。长宽比例适度是至关重要的,那样才能让你的设计与空间配合得天衣无缝。

基本形状
选形和构图

包含要素:

一项设计的版式通常要精心设计,以达到设计的目的、体现设计的功能。

在选择设计的版式时,从实际操作出发必须考虑诸如纸张大小、印刷设备可以印刷的尺寸范围等要素。你必须清楚地知道自己可以掌控的尺寸范围和要达到的设计效果。例如,如果客户需要你先寄送样张,你不至于仅凭想象设计出一件放到哪里都不合适的作品。

经济因素也是选择版式时要考虑到的。不规则图形的制作成本较高,而同样,对纸张大小不合理的裁剪造价也贵,因为它造成了不必要的浪费。

当你考虑这些琐碎因素的同时,可以从审美的角度出发,考察设计的整体感觉。你不会设计出一个导致信息残缺的版式,同样,你也不会让版面大面积空白。

如果你要用照片或插图的话,图片应该是你考虑的重心,因为文字可以很容易地围绕图片排列。如果你处理的仅仅是文字,这可能会很自然地形成一种特殊的格式——例如,如果小说排版的字块中每行字超过10到12个单词,眼睛就很容易疲劳。因此这里首先要考虑的

是确保你选择的形状无损于意思的表达。

相同的情形在你构思设计中其他元素的形状时也会碰到。你需要思考周全到底要达到什么目的,因为形状是设计风格和总体感觉的基础。例如,对象是十几岁孩子的杂志应该有很多振奋人心的形状,并且也可以使版面看上去热闹非凡,但为成人进行的设计就只能是宽阔空间中一个简洁、小型的形状了。

设计效果:

1. 广告牌招贴画一定要与招贴板的大小相一致。2. 正式场合使用的文件焦点在内容,而不是设计,因此,无

论格式还是字块的排版都不能太引人注目。3. 包装的规格大小是根据放置产品的需要确定的。4. 在大街上散发的广告传单必须具备引人注目的形状,才能让人们对它印象深刻。5. 第一印象很重要——合理的版式会强化要传达的信息。6. 超市里的广告册要能够放进顾客的包中,同时还要包含许多信息——长方形并且能够折叠成许多小方块的设计是非常合适的。但重要的是,要确保设计元素不要被折痕掩盖。7. 图片的形状可以根据图片中内容的形状来确定,文字可以排列在图片四周。

8. 如果狭长的图片在水平版面中摆放不下时,也可以采用竖版。

Escape from the city
Balloon tours

2

3

4

Building for Success

The Sky is the Limit

5

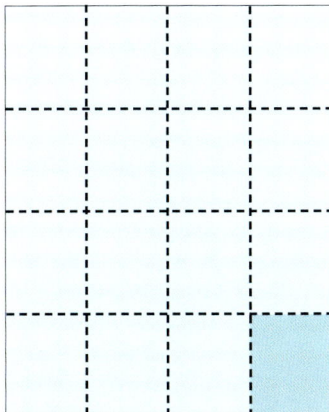

6

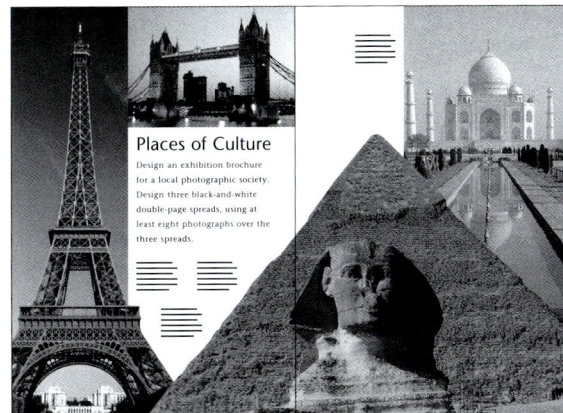

Places of Culture

Design an exhibition brochure for a local photographic society. Design three black-and-white double-page spreads, using at least eight photographs over the three spreads.

7

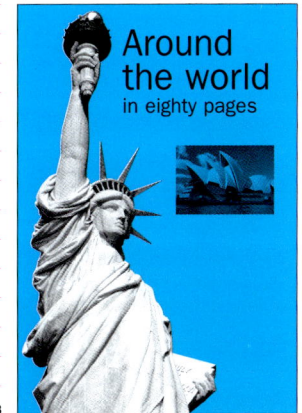

Around the world in eighty pages

8

练习一

- 找一张包含大量细节、能引起关注的图片。
- 浏览图片并选择你认为最适合图片内容的形状格式。
- 尝试将这张图片变成另外一种形状格式。
- 最后，剪切出图片中引人注目的细节部分。审视一下你看到的结果。

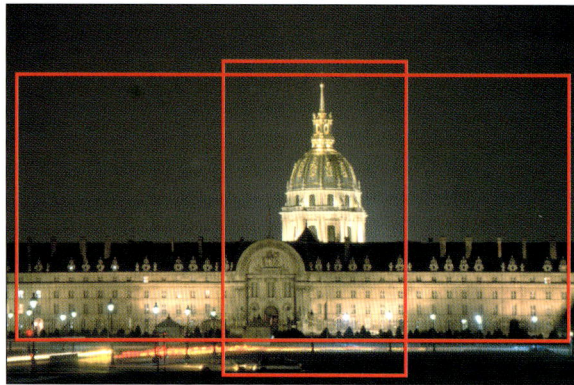

练习二

- 找一首诗、一则菜谱，或者一套说明书。
- 将这首诗、这则菜谱或这套说明书用11磅字体排版并放置在6 x 9英寸的纸张中。
- 为文字配上插图，使说明更清楚。
- 尝试各种排版方式和形状。
- 将你的作品打印出来征求大家的意见。

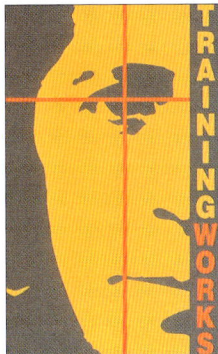

1

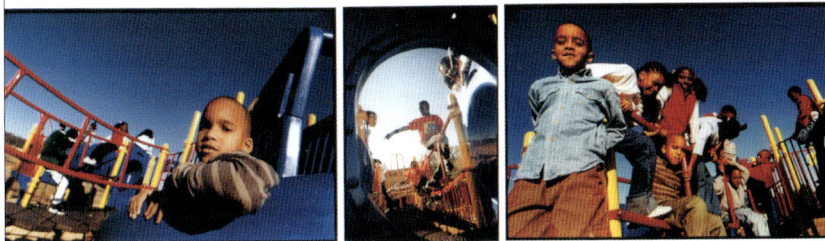

ILENE PERLMAN
PHOTOGRAPHER

2

1. 这张肖像的版式为剪切后的图片提供了恰到好处的空间。

2. 这张风景的版式包含了两张较宽的照片和一张较窄的照片，这是一张为某位摄影师设计的宣传卡片。

CHERRY PIE

Makes 1 pie

Ingredients

2 cups all-purpose flour
1 cup shortening
1/2 cup cold water
1 pinch salt
2 cups pitted sour cherries
1 1/4 cups white sugar
10 teaspoons cornstarch
1 tablespoon butter
1/4 teaspoon almond extract

Directions

1 Cut the shortening into the flour and salt with the whisking blades of a stand mixer until the crumbs are pea sized. Mix in cold war. Refrigerate until chilled through. Roll out dough for a two crust pie. Line a 9 inch pie pan with pastry.

2 Place the cherries, sugar, and cornstarch in a medium size non-aluminum saucepan. Allow the mixture to stand for 10 minutes, or until the cherries are moistened with the sugar. Bring to a boil over medium heat, stirring constantly. Lower the heat; simmer for 1 minute, or until the juices thicken and become translucent. Remove pan from heat, and stir in butter and almond extract. Pour the filling into the pie shell. Cover with top crust.

3 Bake in a preheated 375° F (190° C) oven for 45 to 55 minutes, or until the crust is golden brown.

CHERRY PIE

Makes 1 pie

Directions

1 Cut the shortening into the flour and salt with the whisking blades of a stand mixer until the crumbs are pea sized. Mix in cold war. Refrigerate until chilled through. Roll out dough for a two crust pie. Line a 9 inch pie pan with pastry.

2 Place the cherries, sugar, and cornstarch in a medium size non-aluminum saucepan. Allow the mixture to stand for 10 minutes, or until the cherries are moistened with the sugar. Bring to a boil over medium heat, stirring constantly. Lower the heat; simmer for 1 minute, or until the juices thicken and become translucent. Remove pan from heat, and stir in butter and almond extract. Pour the filling into the pie shell. Cover with top crust.

3 Bake in a preheated 375° F (190° C) oven for 45 to 55 minutes, or until the crust is golden brown.

Ingredients

2 cups all-purpose flour
1 cup shortening
1/2 cup cold water
1 pinch salt
2 cups pitted sour cherries
1 1/4 cups white sugar
10 teaspoons cornstarch
1 tablespoon butter
1/4 teaspoon almond extract

练习三

- 在一张纸上画出 10 到 15 个不同的图形，并把它们剪下来。这些形状应该是可以沿着一边旋转，并且可以随意增减的。
- 按照这些形状从不同的材料上去剪切图形——彩色的纸、闪亮或表面粗糙的纸、报纸或杂志，甚至是织物的边角废料。
- 将剪下的图形放置在一张大卡片上，并按照引人注目的程度进行排列。在每个图形旁边简要注明它的审美价值。

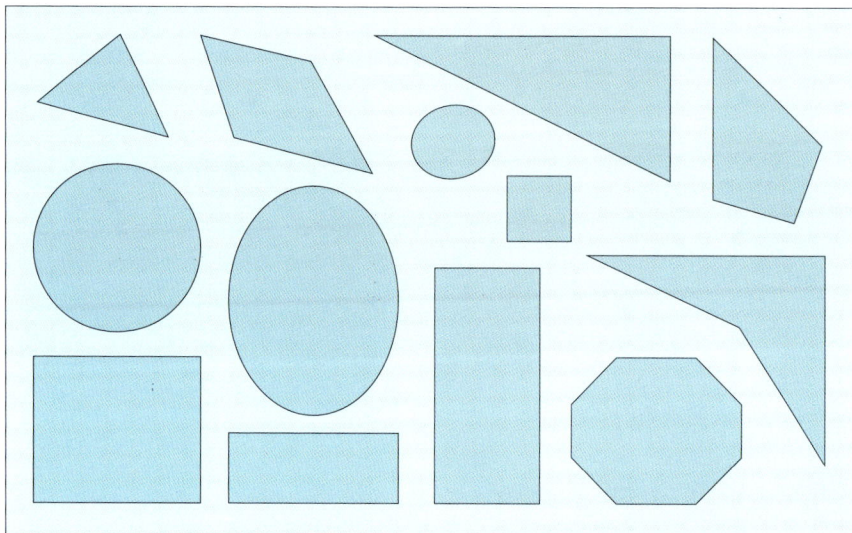

练习四：

- 观察下列单词和设计图形，用电脑或手动将你认为最能够反映单词意思的图形与单词相匹配。单词为：粗糙、柔软、强烈、虚弱、丑陋、单调、规则、令人振奋、普通、流畅、复杂、精致。
- 完成上述练习后，再反过来做一次，选出你认为与单词意思相反的图形。
- 最后，比较这两次练习的结果。第一次练习的结果应该比较称你的心意，而第二次练习的结果就显得有些不协调，并且也不是非常准确。

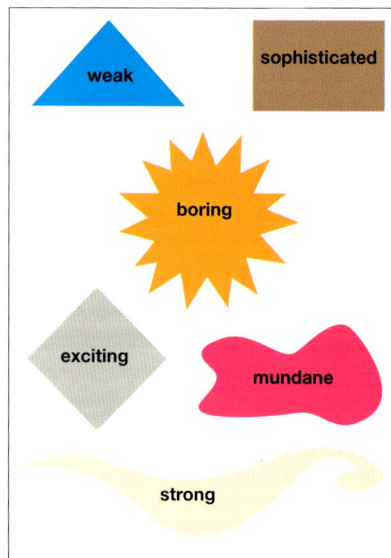

strong mundane

exciting

regular ugly

flowing

weak sophisticated

boring

exciting mundane

strong

1

2

3

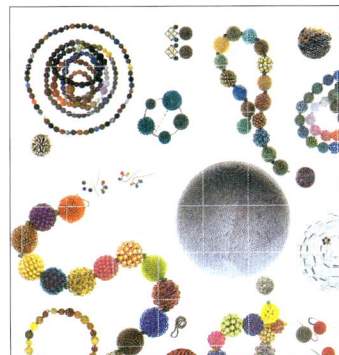

4

1. 这张风景版式为图片和文字提供了足够的空间。为了避免图片或文字被折痕切断，设计采用了各种长度的字块和不同大小的图片。设计效果醒目、亲切，也非常清楚。

2. 这项设计所具有的戏剧性效果是以图片的剪切和字块的特别形状为基础的。

3. 反映面部特写的图片与撕碎的纸片组合在一起，产生了给人感觉强烈的、令人难以忘怀的设计。

4. 散乱地印有连珠的一张邮票——这是一个鼓励使用有创意的个人邮票的广告。

所有优秀设计的各个不同元素之间是平衡的。这种平衡应该会营造出一种协调的感觉，以使读者能够被其吸引，并明白其中的意味。

基本形状
一个单词或一个标题的排版

充分利用空间

在任何一项设计中，准确合理地对各种元素进行排版是建立平衡和协调的关键所在。在第一个练习里，我们将看到在一个页面中对一个单词进行排版的各种方式，并以此来展现单词和它周围空间的各种关系所造成的不同效果。

在一个页面中对一个单词进行排版的方式有很多：可以是左对齐（也就是说每一行的左边对齐）、右对齐或者居中。你也可以改变文字的方向，从水平变为垂直，或者变换角度，甚至打破将文字与页边平行对齐的传统。

设计效果：1. 单词放置在页面中部偏上，产生一种平衡的感觉。2. 单词可以移动到靠近页面顶端——但是可以靠多近呢？3. 单词顶端与页面顶端对齐，造成一种不同寻常的效果。但设计的平衡必须依靠其他元素进行补充。4. 单词放置在页面底部。5. 单词与页面等宽。6. 放大单词。7. 缩小单词。8. 单词靠近页边使页面更有张力。9. 改变单词的方向。10. 不同的角度可能会使单词较难辨认，但却可以形成视觉对比。11. 单词在页面上部排列成半圆形。12. 单词在页面下部排列成半圆形。13. 单词排列成曲线。后三种形式的文字虽然不易辨认，但均具有动态的效果。

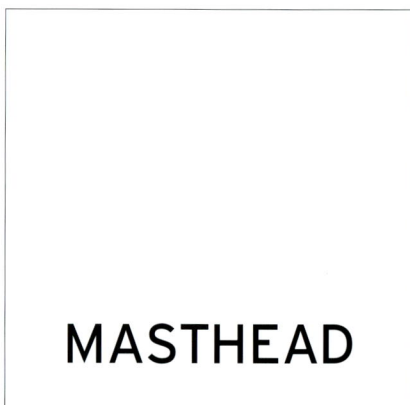

MASTHEAD

MASTHEAD

1

2

MASTHEAD

MASTHEAD

3

4

MASTHEAD

MAST HEAD

MASTHEAD

5

6

7

MASTHEAD

MASTHEAD

MASTHEAD

8

9

10

MASTHEAD

MASTHEAD

MASTHEAD

11

12

13

练习五

- 选择一张 6 × 4 英寸的版面。
- 在版面的可操作区域内区分大小写，打印一个不少于七个字母的单词。
- 将单词居中。
- 设计三种版式变化：左对齐、垂直、倾斜 45 度。
- 尝试将单词排列成半圆形。
- 放大单词排列成半圆形。
- 观察这些设计，并标明它们各造成了什么效果。

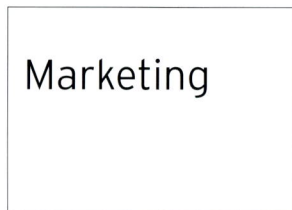

练习六：

- 选择一张 8.5 × 11 英寸的版面。
- 在黑色底面上键入一个至少包含七个字母的单词，区分大小写。
- 字体采用"黑体"，因其在黑色背景中比较清晰易读。
- 尝试各种大小和形状的字体。
- 将版面一分为二，一半白色，一半黑色，并将单词排在其中的一半版面中。
- 完成四种变化后，分析各自产生的效果，并依照你的意见进行修订。

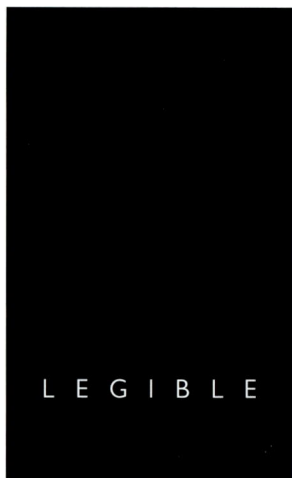

1

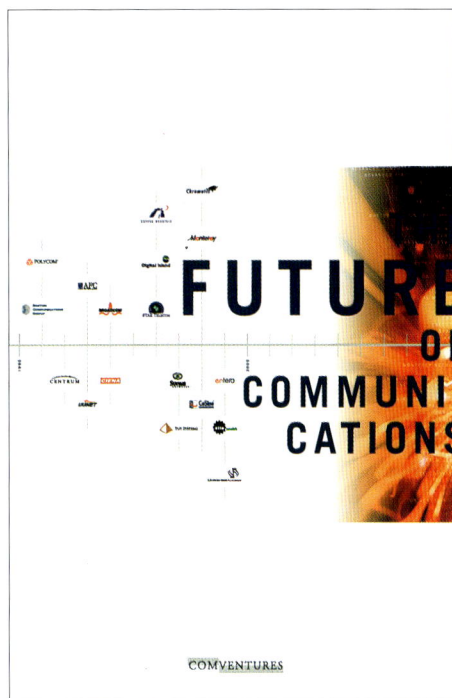

2

1. 标题和相关的细节均被居中，与标志的对称形状形成一体。

2. 字右对齐并排列在紧靠右页边的位置，具有未来派的风格。

3. 水平与垂直字体形成对比的大胆创意，特别合适于以精确和技术为特征的建筑机构。

4. 将标题切割成两部分的设计方式非常有趣，既增加了立体感，又无损于文字的正确辨认。

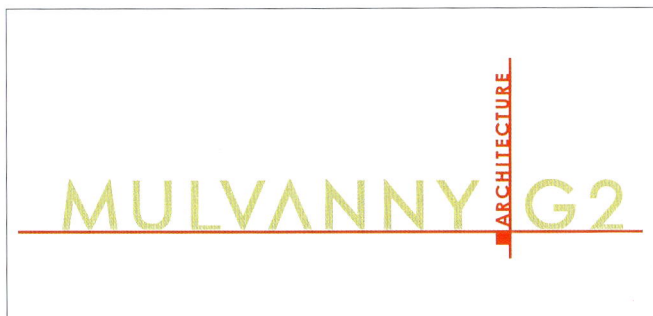

4

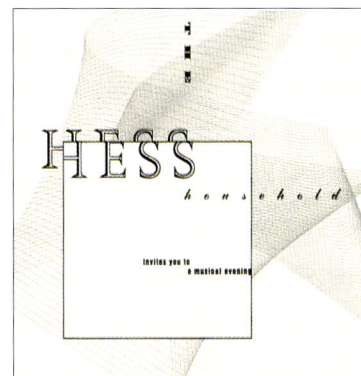

3

当你把标题与图片或一段正文组合起来时，如果要达到预期的效果，标题就不仅要与版面配合，还要与设计中的其他元素相协调。

基本形状
标题与图片或正文字块的组合

利用各种元素创造和谐

设计元素较多的一个好处是你也拥有较多的选择：可以通过改变不同元素的大小、角度和方向使它们在这张版面中能相互协调。

标题与图片或正文的大小不对等会得到比较好的视觉效果。标题大于图片，可以使标题更醒目；标题小于图片，则使图片显得更突出。你必须问自己的一个问题是标题和图片在你的设计中哪一个更重要。

设计效果：标题和正文可以是相同的字体，但大小不同。1. 对称排列。2. 不对称排列。3. 或者可以对标题和图片进行不同程度的突出。图片实际的内容会告诉你以什么为中心。此处强调的是标题。4. 此处强调的是图片。5. 标题与图片所占比例相同又将产生另一种效果。6. 将标题和正文组合起来的排版方式看上去也很有意思。7. 标题的排版采用与正文不同的方向，使页面多了些变化。8. 正文文字也可以依据图片的形状排列。

Typefaces

The ever-increasing flexibility of computer typesetting equipment with the ability to set a wide range of point sizes has done much to popularise the trend towards the use of typeface families. These are typefaces available in a range of weights and derivatives. Many new typefaces now have a semi-bold version that offers the designer a useful choice of weight between the regular and bold designs.

1

Typeface availability

The ever-increasing flexibility of computer typesetting equipment with the ability to set a wide range of point sizes has done much to popularise the trend towards the use of typeface families. These are typefaces available in a range of weights and derivatives. Many new typefaces now have a semi-bold version that offers the designer a useful choice of weight between the regular and bold designs.

2

TYPEFACES

3

Typefaces

 The ever-increasing flexibility of computer typesetting equipment with the ability to set a wide range of point sizes has done much to popularise the trend towards the use of typeface families. These are typefaces available in a range of weights and derivatives. Many new typefaces now have a semi-bold version that offers the designer a useful choice of weight between the regular and bold designs.

6

Typefaces

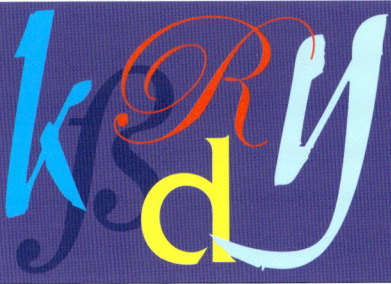

4

Typefaces

The ever-increasing flexibility of computer typesetting equipment with the ability to set a wide range of point sizes has done much to popularise the trend towards the use of typeface families. These are typefaces available in a range of weights and derivatives. Many new typefaces now have a semi-bold version that offers the designer a useful choice of weight between the regular and bold designs.

7

Typefaces

The ever-increasing flexibility of computer typesetting equipment with the ability to set a wide range of point sizes has done much to popularise the trend towards the use of typeface families.

5

Typefaces

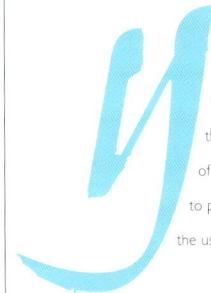

The ever-increasing flexibility of computer typesetting equipment with the ability to set a wide range of point sizes has done much to popularise the trend towards the use of typeface families.

8

练习七

- 从报纸或杂志上剪切一段大约50 到70字的短文，用黑体在6×9英 寸的版面上进行排版，正文对称 页边距，标题在正文上方，并居中 排列。
- 将标题的字体加粗，正文左对齐。 这样出来的效果在页面中不是居 中的。
- 尝试各种大小的标题及标题与正 文的各种排列形式。你可以把标 题放到正文中间，甚至可以让标 题与正文格格不入。不要害怕进 行新的尝试。

Lee Konitz

Associated with 'The birth of the cool', Lee Konitz came from Chicago and has been playing alto sax for over five decades. He is now 75 and has made hundreds of albums, mainly with small record companies. What marks out his career is that he declined early on to be yet another Charlie Parker clone; he has remained an exploratory, innovative saxophonist.

Lee Konitz

Associated with 'The birth of the cool', Lee Konitz came from Chicago and has been playing alto sax for over five decades. He is now 75 and has made hundreds of albums, mainly with small record companies. What marks out his career is that he declined early on to be yet another Charlie Parker clone; he has remained an exploratory, innovative saxophonist.

Lee Konitz Associated with

'The birth of the cool', Lee Konitz came from Chicago and has been playing alto sax for over five decades. He is now 75 and has made hundreds of albums, mainly with small record companies. What marks out his career is that he declined early on to be yet another Charlie Parker clone; he has remained an exploratory, innovative saxophonist.

Lee Konitz

Associated with 'The birth of the cool', Lee Konitz came from Chicago and has been playing alto sax for over five decades. He is now 75 and has made hundreds of albums, mainly with small record companies. What marks out his career is that he declined early on to be yet another Charlie Parker clone; he has remained an exploratory, innovative saxophonist.

练习八

- 选择一张照片并与文字组合，图 片应占据80%的版面。
- 改变图片和文字的比例，图片占 据20%的版面，文字和空白部分 占据80%。也可以对图片和文字 的摆放角度进行变换。
- 最后，挑选出你自己认为比较好 的图片和文字的组合。
- 完成练习后，分析各种方式的效 果，并根据你自己的意见进行修 改。

two magpies = 雙喜 double

happiness (*shuāng xǐ*)

magpie / The characters for 'magpie,' *xǐ què*, literally mean the 'bird of happiness.' A picture of two magpies facing each other stands for 'double happiness,' *shuāng xǐ*, symbolic of conjugal bliss. The call of a magpie foretells the arrival of a guest, good news, or good fortune. A magpie resting on a plum branch conveys the wish 'happiness before one's brow,' *xǐ shàng méi shāo*, as the word for 'plum' and 'brow' are both pronounced *méi*. Magpies also served to preserve the integrity of a marriage, according to legend. When a husband and wife were to be apart for any reason, they would break a mirror and each take half. If the wife was unfaithful, her half of the mirror turned into a magpie that flew back and informed her husband. Consequently, an image of a magpie is often placed on the back of a mirror.

double happiness 🈲 171

1

Il y en a que j'aime plus que d'autres. Les gens ne les comprennent pas tous de la même façon: quand ils les découvrent, je les découvre. Je suis étonné et bouleversé quand, parfois, j'en emmène un au format gigantesque de l'affiche.
Leur petite taille, pratique et humble, impose une grande précision dans la narration, par le mot, ou l'image ; une gymnastique que j'adore, celle du raccourci.
Il y a des "collectors", des "petits coupons" qui ont existé et que j'ai supprimé par manque de conviction. D'autres que j'ai réimprimé, avec des corrections.
Il y a des collectionneurs. Il y a eu des articles dans les journaux sur ces petites images. L'un se promènent à Paris, à Londres, ou à New York.

Cette exposition s'appelle "Super !".
Le "SUPER !"-VERNISSAGE a eu lieu le 26 mai 00. L'exposition dure jusqu'au 29 septembre 00, et sera peut-être remont(r)ée à Paris. D'ordinaire données, ces petites images seront vendues 1 F pièce pendant toute la durée de l'exposition, dans le "SUPER !"-MARCHÉ.

Cinq premiers "petits coupons" créés en 1994.
La "collection" en compte aujourd'hui, une centaine.
Cent prochains sont à l'état de notes, et je suis aujourd'hui certain d'avoir le profond désir d'en créer de nouveaux, toute ma vie durant.

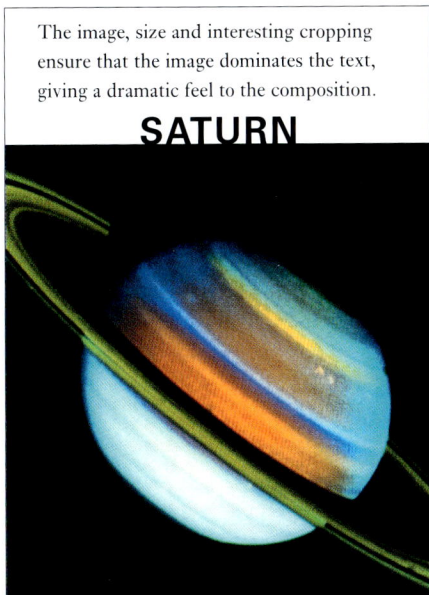

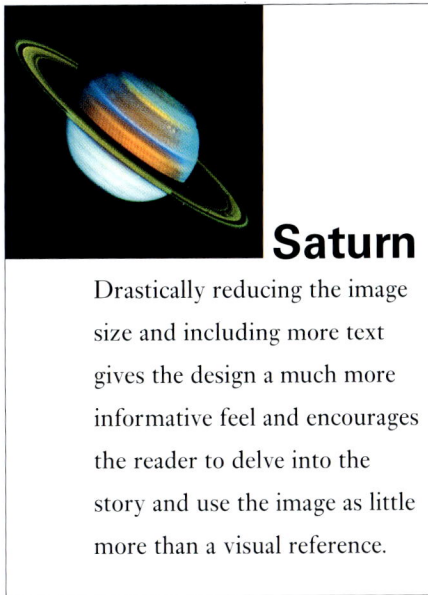

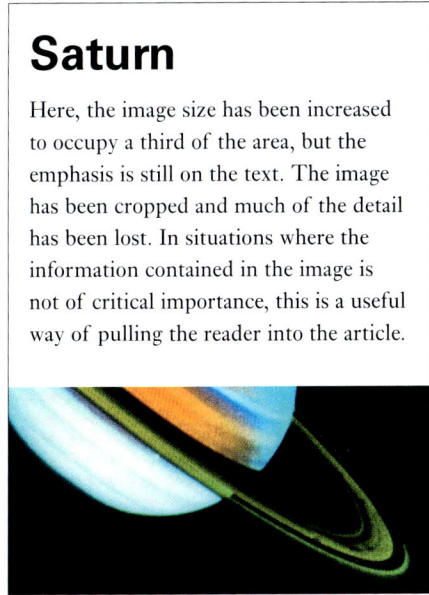

2

1. 这项设计的成功之处在于解决了将两种完全不同的文字组合在一起的难题。这两种文字的排版方式使得它们都易于辨认且不互相冲突。

2. 采用有颜色的标题意味着可以将它覆盖在正文上，形成更强烈的视觉冲击。

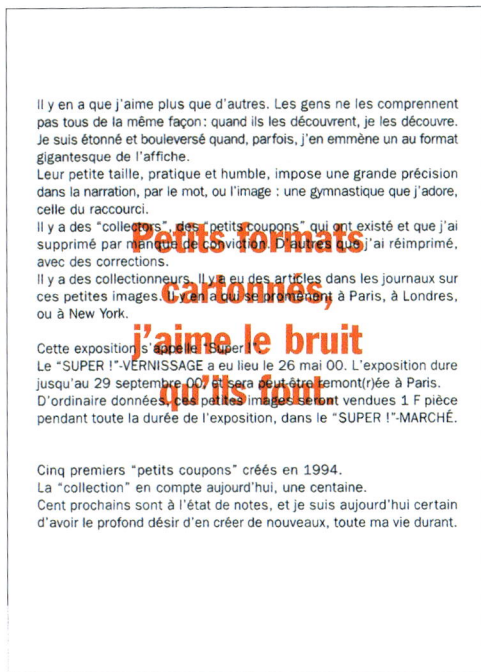

The image, size and interesting cropping ensure that the image dominates the text, giving a dramatic feel to the composition.

SATURN

Saturn

Drastically reducing the image size and including more text gives the design a much more informative feel and encourages the reader to delve into the story and use the image as little more than a visual reference.

Saturn

Here, the image size has been increased to occupy a third of the area, but the emphasis is still on the text. The image has been cropped and much of the detail has been lost. In situations where the information contained in the image is not of critical importance, this is a useful way of pulling the reader into the article.

在一项设计中，你不得不处理一些不同的元素，诸如两行或者三行标题、一个副标题、正文及其插图。你考虑的元素越多，取得协调就越困难。

平衡和协调
多种元素的组合与排列

处理多种元素

第一步是权衡每一元素的重要性，确定你要给每个元素多大程度的强调。随后从选择标题的字体、大小、宽度着手。再以相同的程序处理副标题和正文。

正文的排版也要注意。正文应占多少行？标题与正文的位置如何摆放？图片与文字的宽度是否匹配？或者版面看上去是否太挤或太空？在你作出最后决定前，尝试各种排版方式。

设计效果：1. 对称的排版方式。文字和图片各占版面的一半，标题位于文字和图片上方正中的位置。这样的设计透出的是一种平静的和谐。2. 不对称的排版方式，与前一种相比，文字占的地方较少，重点在图片上。3. 标题是版面中的重点。4. 图片在版面中居突出地位。5. 图片仍旧占据主导地位，但排版的格式不同。6. 标题的方向与正文方向不一致。7. 排版形状发生变化，标题、副标题、正文和图片组合在一起。8. 栏中有空白的多列式排版方式，突出了各栏标题。

Heading

Here, text and image take up equal amounts of space. The shape of the image allows more background detail to be included, thus reducing the emphasis on the main subject, the balloon. Justifying the text and centring both image and text gives the design a rather static appearance. The overall composition is gentle in feel.

1

Heading

In this example the image has much more impact, even though the image and text have the same proportions as in the first example. The reason for this is that the background detail has been reduced, drawing attention much more quickly to the balloon. The text is ranged left and the overall appearance has more drama than the one above.

2

HEADING

This approach uses a strong display heading to catch the reader's eye before he or she moves on to the image and text. The size of the heading in bold capitals ensures that it dominates the design. The other elements in the arrangement have been reduced to the point where they become almost insignificant.

3

The image has been removed completely, and so the reader has to be attracted into the subject by the display heading. The heading could be set very large or, if space is tight, given extra emphasis by being placed in vertically. In this example, there are four text columns. With columns this narrow, it is better to range the type left to avoid excessive word spacing and word breaks. If there is sufficient copy, you could introduce some subheadings. These can be set in a bold version of the typeface.

Heading

6

Heading

Here the image size is enlarged to occupy two-thirds of the format, giving a change of emphasis. This has a similar effect to that shown in the second example, in that the image is the main attraction and text is pushed into the background.

4

Heading

Subhead Here there is a change of direction of image, heading and text. By slicing away parts of the image, the reflection of the balloon has gone. This arrangement is much more dynamic, but the text is less readable and there is not nearly as much information in the image.

7

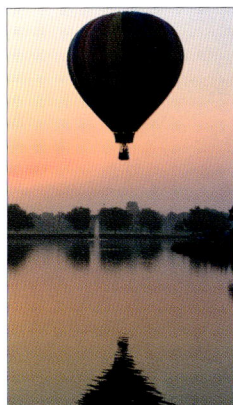

Heading

The emphasis here remains on the image, although text and image occupy more or less the same amount of space. Cropping the image and concentrating upon the balloon quickly attracts the eye to the subject. This treatment means losing background detail.

5

Heading 1

This example gives the best of both worlds. The image attracts attention; at the same time, the three columns with their separate headings give

Heading 2

added visual interest. This type of composition can be flexible if copy is limited; as a variation, you could remove one of the columns

Heading 3

and increase the image size or the column measure. As in example 7, range the text left without any word breaks.

8

练习九

- 采用与练习八中相同的版面格式和风格，将正文和照片并排放在一起，标题位于上方正中的位置。
- 改变排版方式，图片占据页面的主导地位，文字和标题不居中。

Equal Proportions

The image size and the text have equal emphasis. 'Bleeding off' the image into the left-hand edge gives the design added visual impact.

The problem here is that a lot of the image area has now been lost, and so its informational value is diminished.

Larger Image

This arrangement has even more drama. The image size has been increased and now bleeds off on two sides. Again, the problem is the information that is lost as the image is decreased further: the rings of the planet have almost disappeared.

练习十

- 采用与练习九中相同的素材，把图片拉窄拉长。在原来的基础上再加上些文字，并将标题改变方向。
- 最后，在图片巾剪切出圆形图案，并将文字围绕图片紧密排列。
- 完成练习后，分析各种排版方式的效果，并根据你的修改意见对它们进行修正。

Abstraction is Nearer

Visually, this looks more interesting because the proportions of the two elements are unequal. An added visual trick is that the heading runs at an angle. By restricting the image to a small area, it becomes more abstract – but the amount of information in the image has been dramatically reduced.

Circular Shapes

Although the image shape looks good, it has no informational value whatsoever and has become pure abstraction. This is acceptable if shape alone is the criteria for the design. The circular shape invites a text to be set around it, which gives a good structure to the composition. In this type of arrangement, the text should be ranged left; this avoids the kind of exaggerated word spaces that would be produced with justified setting.

多种元素的组合与排列

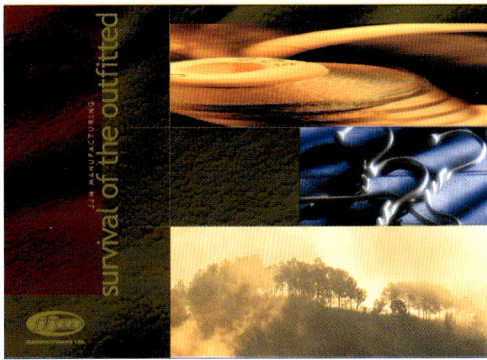

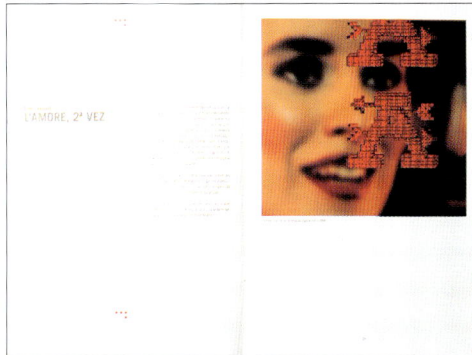

1. 垂直放置的标题与风景图片
   完美地结合起来，风景图片
   也有足够的宽度展开。
2. 图片和文字的底端对齐。这
   样的组合留出了大量空间，
   使设计达到平衡并具有现代
   的效果。
3. 分两行的标题和副标题，其
   摆放位置与图片的风格配合
   得天衣无缝。
4. 明显的自由风格的设计，图
   片和文字的排版方式为这项
   时髦的设计增加了张力和可
   看性。
5. 用圆形进行排版是非同寻常
   的尝试。文字沿着圆形的弧
   线排列，使得设计引人注目
   且和谐。

网格将页面划分成一个个小方块或单元，这些单元是对文字、图片和说明进行排版的参照。网格扮演着"组织者"的角色，特别是当设计师们为了一项设计而协同合作时，它更能在按照顺序排列元素上帮设计师大忙，而且也是一个重要的工具。

平衡和协调
网格的使用

内容限定网格

网格中单元格的数量根据设计素材的复杂性确定。设计内容越多，网格中单元格的数量也就越多。

一种简单的网格是将页面等分，水平 3 个单元格，垂直 3 个单元格。

更复杂的网格可能每个单元格的大小不同，且单元格的数量也较多，使之能够容纳各种长度的字块和各种形状的图片。

设计效果： 网格要留出合适的页边，每个单元格之间也要有一条空白的分割带。1. 简单的网格包含九个单元格。先标出上下左右的页边，在设计网格时，网格内采用细线，而网格边缘采用1磅的线条。2. 如图所示，设计元素应从左上部的第一个单元格开始排列。3. 设计你的网格时，显示出打开的两个页面是至关重要的，因为我们在阅读时总是同时看到打开的两个页面。4. 图片应占据整个单元格。5. 图片也可以占用不止一个单元格的空间，如 1 × 1、1 × 3、2 × 2、3 × 3 个单元格。6. 具有更多单元格且单元格不相等的更复杂的网格。7. 狭长的单元格可以用来放置图片的说明文字。

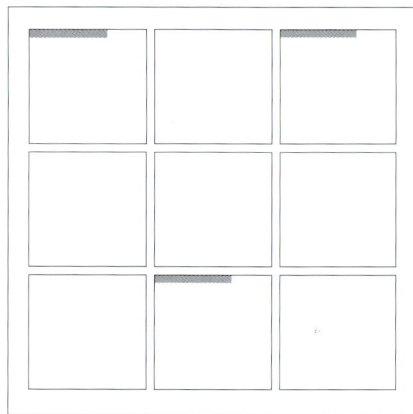

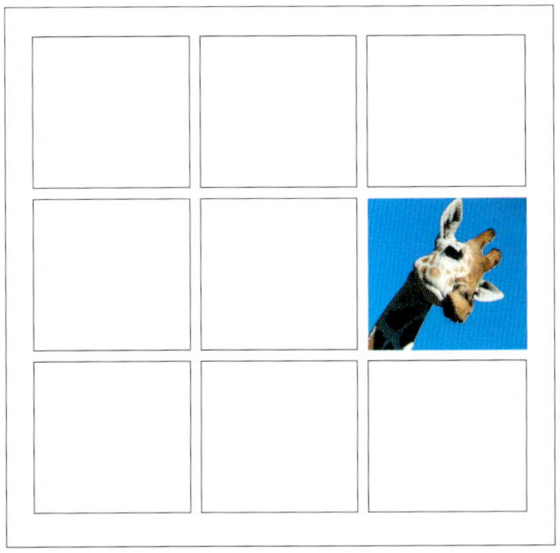

4

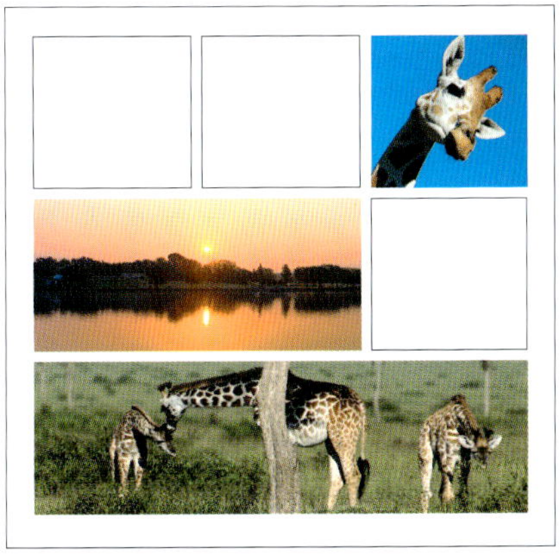

5

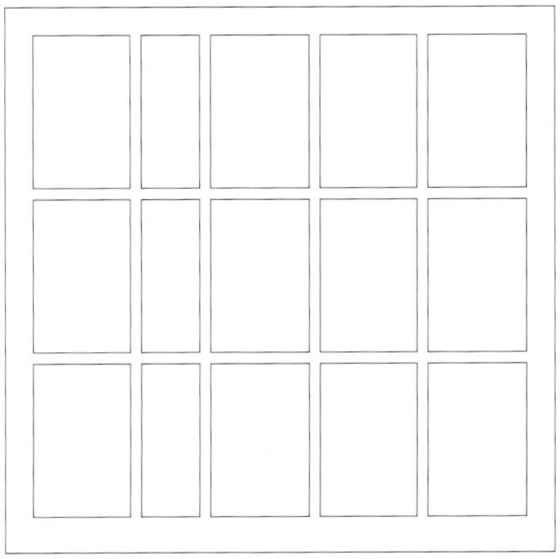

6

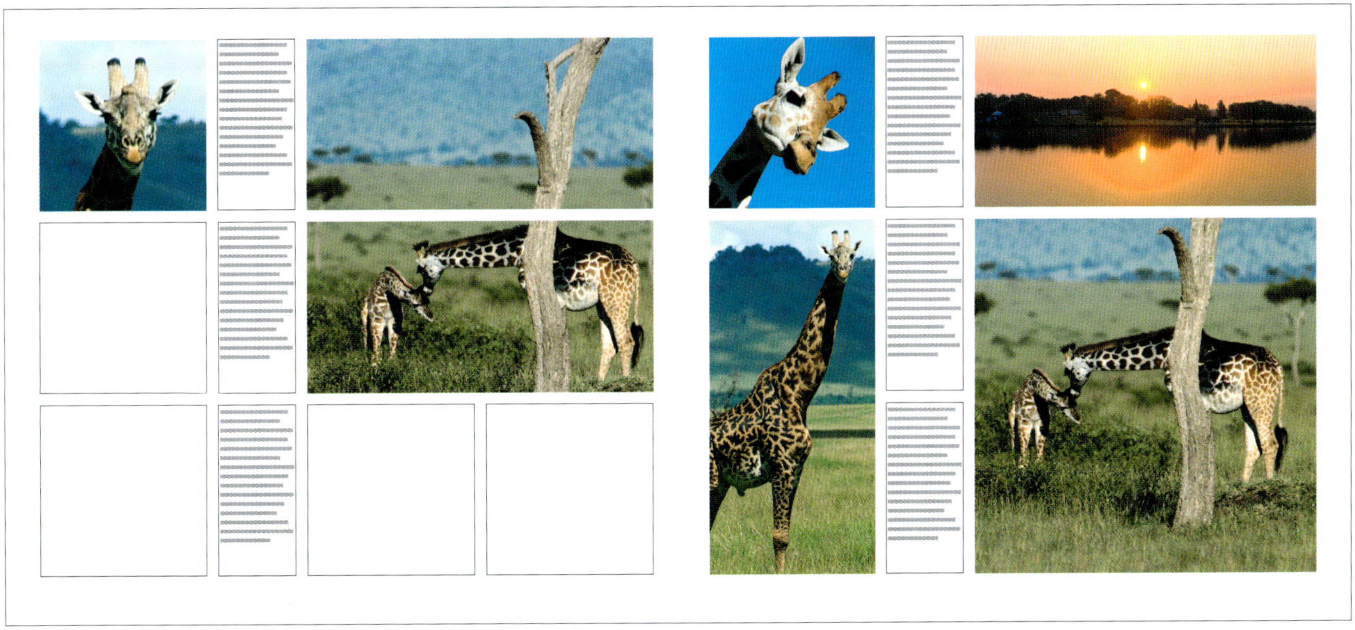

7

练习十一

- 选择一张 9 × 6 英寸的页面，将页面等分成 9 个单元格，留出 0.25 英寸的页边。
- 使用网格进行排版。用无衬线的铅字体 10 磅键入一段文字，并占用 5 个单元格，再加上标题，确保你的文字从左上部的第一个单元格开始排列。
- 在网格中的不同单元格进行以上练习。

Heading

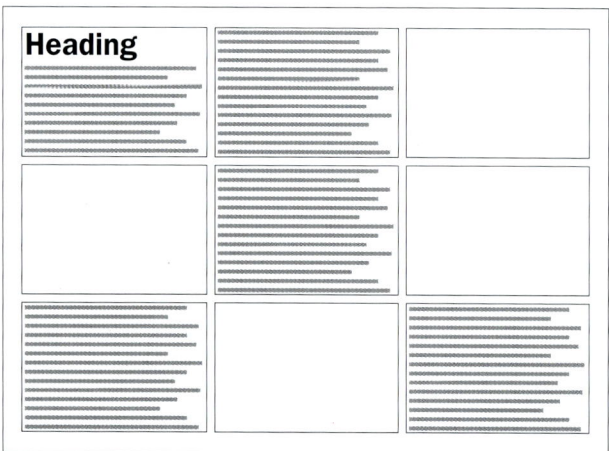

练习十二

- 将页面大小调整为 8.5 × 7.75 英寸，构建一个网格，网格的垂直方向分为平均的四个单元格，水平方向分为五个单元格，其中的一个单元格宽度是其他四个单元格的一半。
- 从杂志上选择一段文字，采用 9 磅字体，打头字母用 2 磅字体，并左对齐。所有的文字不要超出一个单元格的范围。
- 选择三幅图片，采用与练习十一中相同的标题，练习各种排版方式：改变图片的大小、改变正文字块的大小，或者改变占用单元格的数量。在改变正文字块尺寸时，记住将文字控制在单元格范围内。

Heading

Heading

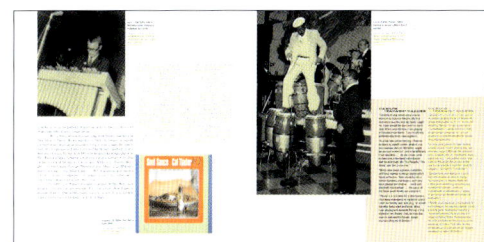

1. 这张标题页面合理的利用了网格，页面水平、垂直方向的对齐方式简洁明了。

2–3. 这两张展开的页面显示了网格在设计中的变化。正文字块的尺寸富有变化，而图片的说明文字只占用一个单元格的宽度。这样的设计使页面丰富、活泼、充满激情。

字体分为正文字体和标题字体两类。这两类字体的区别在于字号的大小。正文字体用在文章正文部分，通常字号为8到12磅。标题字体大于14磅，通常用在标题或正文前的简介中。

字体
标题字体

选择一种标题字体

从某种程度上说，在选择一种标题字体时确实要"挑花眼"：有上千种字体可以供你挑选。在你做选择时，有几条建议可供参考：

字体的视觉效果如何？一些字体看上去优雅柔和，另一些却强烈抢眼。你想让字体反映出正文内容吗？利用字体的特征及字体产生的效果。你想让标题字体与正文相协调（也即是说与正文的字体一致）还是形成对比（例如正文用有衬线的字体而标题用无衬线的字体或黑体）？

设计效果： 1. 有衬线字体的特征是笔画末端有一条短的衬线。2. 无衬线字体笔画末端没有短的衬线。3. times字体是常用的有衬线的字体，给人以传统的感觉。4. 将stone sans具有现代风格的字体与3中展示的times的传统字体相比较。5. 这里展示的是较为纤细的有衬线的字体，例如波多尼字体。6. 衬线方正的字体，如罗克威尔字体。7. 圆弧衬线字体，如century schoolbook。8. 当各种不同风格的字体放在一起时，你可以看出它们的不同之处。9–13. 字体的视觉效果与正文内容相匹配的范例。

A serif
typeface

1

A sans serif
typeface

2

The tradition
of a
classic serif

3

The
modernity
of a
clean, sharp
sans serif

4

Bodoni
has a
hairline serif

Rockwell
has a
slab serif

Century
has a
slab bracket

bracket
hairline
slab bracket
slab

The
mechanical
achievement
of Courier

The soft
approach of
Bembo

The
AUTHORITY
of Gill

The delicacy of
Copperplate Script

The
power
of
Helvetica

练习十三

- 选择一张 8.5 × 11 英寸的
  页面，并为下列文字片断
  选择与其意思相协调的标
  题字体，大小写不限。
  - 纯洁的诗歌
  - 冲突的痕迹
  - 勿踏草坪
  - 沿着记忆的小径

pure poetry

traces of
conflict

KEEP OFF
THE GRASS

Down Memory Lane

练习十四

- 在这个练习中，你要换一
  种思路，字体成为设计的
  主导。用下列字体排版，键
  入与字体特征相符的文字。
  - bembo 斜体
  - helvetica 粗休
  - rockwell 细体

Hand crafted
porcelain since 1851.

You will find a lifejacket
located underneath
your seat.

October 18, 23.00 pm.
Suspect enters side
entrance. Visibility is poor.

Uyl en Spiegel

1. 粗糙的、手工制作般的油印字体与图片结合
   得完美无缺。
2. 采用改变字体大小、对齐方式及放置位置等
   手段将字体与文字含义配合是非常聪明的处
   理方法。字母间距的缩小也营造出了紧张的
   感觉。
3. 字体被极富想象地设计成模仿跳舞的感觉。
4. 无衬线字体的使用，并拉伸开来，使设计具
   有一种扩张的感觉，与单词的意思吻合。

2

3

4

有某些字体可以称之为"经典"。这些正文字体经受住了时间的考验且被广泛使用,原因在于其本身在形式和审美上所具有的特质。

字体
正文与字体

正文的字体

诸如bembo、garamond、caslon、times、palatino及plantin等字体均可称之为经典字体,当它们与其他标准的字体一起使用时,也可以保证其易读性。

在选择正文字体时要考虑的因素与选择标题字体时要考虑的差不多,但有一条是必须首先考虑的:字体的功能。换句话说,正文的意图何在?它只是像小说一样让人们能够连续读下去吗?或者它们被小标题分割成若干部分,让读者能够轻松地阅读每一小标题下的内容?或者字块的整体感觉比实际含义更重要?

在选择正文字体时,有如下因素要考虑:

● 字体的 x 坐标。

● 字体的宽度。

● 字体的视觉效果;它可以影响并强化文字的含义。

设计效果: 1. 字体的x坐标指的是字体的主体高度(由x的小写字母而得名)。上部分指任何超过x坐标的部分(如小写字母b的上半部分)。下部分则指任何低于x坐标的部分(如小写字母p)。2. 在相同的磅值上,x坐标较低的字体(如bembo)比x坐标较高的字体(如times new roman)看上去小一些。3. 比较宽的字体(如century schoolbook)占用的空间比较窄的字体大(如ehrhardt)。4-6. 建议类文字:命令或警告类的文字最好采用无衬线的字体,因其看上去比经典的衬线字体或斜体更具权威性。7. 连续不断的文字:前面提到的经典的字体均是很好的选择。每行最合适的单词数目是10到12个。8. 事实陈述类文字:这类文字仍然需要能够被连续不断地读取,但当中被醒目的大标题和小标题隔断。此时无衬线的字体较为合适,因其具有设计精良、宽度合适的特点。9-11、杂志文章:这里首先要考虑的不是易读性,而是时尚感。特别是时尚类杂志,需要更具些冒险性。经典的字体此时看上去太柔和,而诸如erhardt、stone和rotis类的字体较为合适。

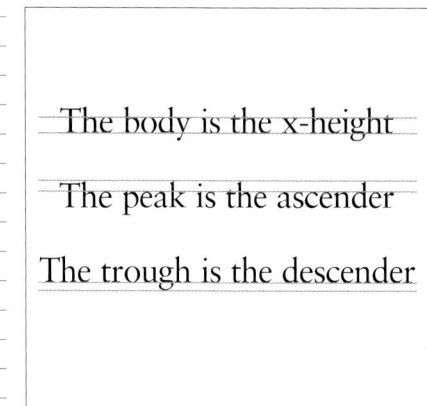

The body is the x-height

The peak is the ascender

The trough is the descender

1

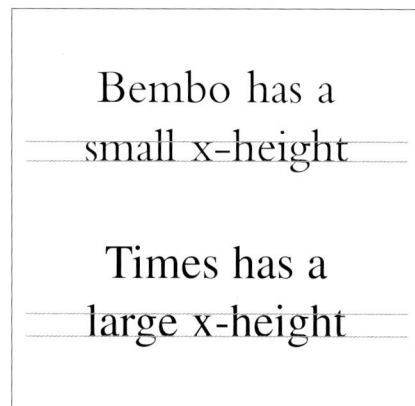

Bembo has a small x-height

Times has a large x-height

2

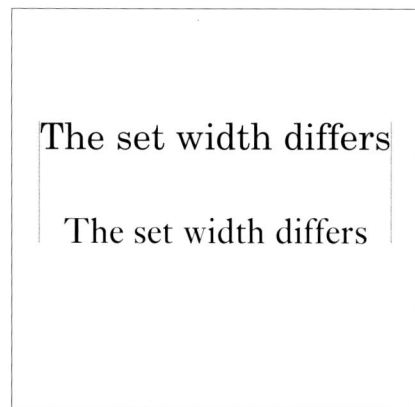

The set width differs

The set width differs

3

Keep off the grass

4

Keep off the grass

5

Keep off the grass

6

Mackintosh's influence on the avant-garde abroad was great, especially in Germany and Austria, so much so that the advanced style of the early 20th century was sometimes known as 'Mackintoshismus'. His work was exhibited in Budapest, Munich, Dresden, Venice and Moscow, arousing interest and excitement everywhere. From 1914 he lived in London and Port Vendres. Thereafter, apart from a house in Northampton, none of his major architectural projects reached the stage of execution, although he did complete some work as a designer of fabrics, book covers, furniture and painted watercolours.

7

Helvetica
Ubiquitous sans serif typeface designed by Max Meidinger and Edouard Hoffman and issued by the Swiss typefoundry, Hass, in 1957.

Origin
It is based on Akzidenz Grotesque, an alphabet popular at the turn of the century.

8

Ehrhardt
This typeface is based on originals from the Ehrhardt foundry from Leipzig in the early eighteenth century. An elegant serif typeface similar to Jenson, it has a narrow set width that allows more characters to the line, making it ideal for magazine work.

9

Stone Sans
This versatile sans serif was designed by Sumner Stone in 1987 to meet the requirements of low-resolution laser printing. There are three weights available – medium, semi-bold and bold. The face harmonises with Stone Serif and Stone Informal.

10

Rotis Semi-Sans
A popular sans serif throughout the 1990s, Rotis Semi-Sans was designed by Otl Aicher in 1989. Condensed in structure, some characters exhibit variations in the thickness of the strokes. It is available in four weights – light, regular, medium and bold.

11

练习十五

- 选择一张 4 英寸左右的版面，键入至少 4 行文字，然后分别用 4 磅、8 磅、12 磅和 16 磅的 sabon 字体进行排版。看看哪种大小的字体在这样大小的版面中最合适。
- 选择合适的字号，并用 century schoolbook 字体（较宽的字体）排版，再用较窄的字体，如 ehrhardt 重复上述练习。
- 比较两者的效果。

There are many factors to consider when thinking about readability and legibility of the text. These include the line length (measure), the size of type, the weight of type, the amount of leading and the size of the x-height.

4pt

There are many factors to consider when thinking about readability and legibility of the text. These include the line length (measure), the size of type, the weight of type, the amount of leading and the size of the x-height.

8pt

There are many factors to consider when thinking about readability and legibility of the text. These include the line length (measure), the size of type, the weight of type, the amount of leading and the size of the x-height.

10pt

There are many factors to consider when thinking about readability and legibility of the text. These include the line length (measure), the size of type, the weight of type, the amount of leading and the size of the x-height.

12pt

There are many factors to consider when thinking about readability and legibility of the text. These include the line length (measure), the size of type, the weight of type, the amount of leading and the size of the x-height.

16pt

练习十六

- 用 12 磅 bembo 字体键入如下内容："The apparent size of a typeface varies according to the x-height."
- 用 Times New Roman 字体重复练习。
- 比较两者的效果。

The apparent size of a typeface varies according to the x-height.

The apparent size of a typeface varies according to the x-height.

[X I N E T]

VERSATILE

WebNative and

WebNative Venture allow users to access the

FullPress server over the Internet. Regardless of their location,

users can search for assets online or in archives. Everything that

WebNative displays to users comes from the heart of the FullPress server. As

soon as a file is put on the server, it is available for use. If a file is modified, the

changes are automatically made apparent through WebNative and WebNative Venture.

Unlike some systems that sit on the outside looking in, Xinet's digital asset

management solution operates right in the middle of the workflow. Because Xinet

software is based around the filesystem, it is completely and automatically

integrated into production. There is no one else optimizing workflows

and managing assets like Xinet: no one else works

so closely with the filesystem.

"Now our server produces
twice as much due to
FullPress' efficiency and
increased business from
customers. The workflow
is definitely faster because
we don't have to double-
check our work like we
did before."

CLAUS KOLB
Chief Executive Officer
Kolb Digital GmbH
(digital and commercial printer)
Munich, Germany

1

[Some level considerations on the history of the poster]

É rica e densa de acontecimentos a já longa história do cartaz, produto considerado, ao menos ao longo do século XX, como elemento fundamental de comunicação, sobretudo ligado ao desenvolvimento e progresso das cidades e, mais em geral, da vida urbana. É, a este respeito, muito significativo que, desde muito cedo, apareçam referências à presença de cartazes como elementos decisivos de comunicação. Refira-se, a título de mero mas muito esclarecedor exemplo, que já o Daniel Defoe, no seu célebre Diário da Peste de Londres, descrevia há mais de duzentos anos a sua cidade, testemunhando que esta se encontrava coberta de cartazes. Sem precisar de retroceder até à época fundadora de Gutenberg ou sem procurar estudar, nos seus aspectos semiológicos ou comunicacionais, as muitas formas que o cartaz foi tomando até se volver nesta linguagem quase específica que se tornou nossa contemporânea, é preciso lembrar, ao menos, a sua época dourada, em que muitos foram os artistas que enveredaram pela sua execução, assumindo-o como veículo ideal de

The history of the poster is long, rich and eventful. It is a product considered, throughout the twentieth century at least, a fundamental component of the modern means of communication, especially in its links with the growth and development of cities and, in more general terms, the urban way of life. In this respect it is highly significant that references to the presence of posters as decisive factors in methods of communication appear from a very early date. For a colorful but very revealing example or two, take a look at Daniel Defoe's famous Journal of the Plague Year, where he was already describing London as being covered

3.º CONGRESSO INTERNACIONAL DE ESTUDOS PESSOANOS / 19 Centro de Estudos Pessoanos

2

2

1. 正文部分的衬线字体和标题的无衬线字体搭配协调，将可读性与精致性结合了起来。宽阔的行间距和较短的行列使传统的衬线字体显得非常清晰。标题字体利用了圆形排列，下半圆反射式的字体增加了设计风格的趣味性。视觉效果是至关重要的，因此对正文的编辑也非常要紧，因为这样可以避免错误的断行或单词间的距离过宽，而这恰恰又可能破坏圆形设计的效果。

2. 这是个有趣的组合：一个字块的尺寸特别宽，另一个字块则是常见的宽度。超宽的设计只有在文字数量有限时才能成功地发挥作用，例如在这项设计中，否则正文读起来会比较困难。

你可以通过改变字体的磅值和风格这两个主要方面而使版面焕然一新。许多字体都有相对立的粗体，因此改变字体的磅值很容易，另一种影响视觉效果的调整方式是改变行间距和字间距。

版面设置
磅值、风格、行间距和字间距

参考如下建议：

改变磅值的最好方式是使用无衬线的字体。几乎所有无衬线的字体都有细体、正常体、粗体和特粗体。有衬线的字体也有相应的粗体，但在许多设计中，它们往往过于粗黑，损害了设计原来具有的美感。

改变字体风格时，要避免使用同一类的字体，如 Bembo 和 Garamond。

行间距的大小（正文行与行之间的距离）依据字块大小、x坐标和字体的磅值而定。因此排版需要不断调整以使每行字体能够以相同的视觉效果结合在一起。这种设置依据的是x坐标上部分和下部分的组合情况。

字间距意味着字母间距离的调整。某些字母组合在一起时往往形成较难处置的字间距。

设计效果： 1. 改变字体的磅值可以有很多选择。这里展示的是Helvetica字体的细体、正常体和粗体。2. 风格对立的字体组合在一起也能收到很好的效果（如正文用有衬线的字体，标题用无衬线的字体）。3. 将相同类别的字体混合在一起（即将两种有衬线的字体或两种无衬线的字体组合）会产生冲突。这里是两种不同时期的无衬线的字体，Futura体和Helvetica体。

4. 在排版时，你选择的行间距的大小依照字母间的相互关系确定。5. 不合适的行间距，换行的磅值小于字体的磅值，虽然看上去有趣，但却不易读取。6-7. 在设置正文时，x坐标影响着行间距。大多数无衬线的字体x坐标较大，因此行间距比x坐标较小的字体宽。8. 行内间距调整功能可以显著调节所有字符间在同一行内的距离。9-10. 字间距调整功能调节的是两个字符间的距离。有比较长的斜向笔画的字符（如 A、V、W、Y）占用的空间比其他字符大，需要通过改变字间距来补偿。11. 连字也是印刷中常采用的方式，某些字符被组合起来形成一个字符。大多数字体都包含有连字，如字母f后紧跟字母i，字母f后紧跟字母l等，因为这些字母放在一起时容易组合。

Changing weight gives options

Changing weight gives options

Changing weight gives options

1

Ranged-left arrangements

If you want consistent, uniform word spacing, then a ranged-left setting style is necessary. However, if the measure is too short and the type size too big, there will be unsightly space at the end of many of the lines. Between paragraphs in this style of setting you could insert a half-line or full-line space; alternatively, you could indent the first line of each paragraph by 1 cm. You should also think about the amount of leading: remember that the readability of all typefaces is improved with leading and as a matter of course you should not use the auto setting on the computer. Always specify a specific point size for the leading.

2

Helvetica

This is Futura. Don't mix the two.

Sans serifs like Franklin Gothic have large x-heights and need generous leading.

Kerning is a useful tool for controlling the space between c h a r a c t e r s.

The best designers use their eyes and trust their judgement.

Serifs like Bembo have small x-heights so there is no real need for generous leading.

AV   VA
WY   YW
AW   WA
AY   YA
VW   YV

The best designers use their eyes and trust their judgement.

Tracking at -10

Tracking at 10

Tracking at 30

T r a c k i n g   a t   5 0

fi   fl

fi   fl

3

4

5

6

7

8

9

10

11

练习十七

- 选择一张 8.5 × 11 英寸的版面。
- 用无衬线字体键入"You do not have to shout to be heard",区分大小写。
- 选择与版面大小相符的字体大小换行恰当,不要将单词断开。
- 改变字体的磅值反映出你对这句话的理解。
- 采用不合适的行间距,使字母紧挨在一起,最后甚至重叠在一起。
- 设置如下文字:"The Language and Culture of Cartography in the Renaissance",分三行,并用与版面合适的字体,行距固定(即没有行间距)。
- 加入行间距,并改变字间距,直至达到平衡。

You do not have to shout to be heard

You do not have to **shout** to be heard

You do not have to shout to be heard

练习十八

- 选择有衬线的字体用大写键入如下单词:"VARIOUS WAYS OF DRAWING WAVES"。观察单词和字母间的距离。
- 调整字间距,平衡占用空间太大或太小的字母,A、V 和 W 通常是产生问题的字母。
- 调整行内字间距,使文字与版面匹配。
- 将调整过字间距和行内字间距的版面与最初的版面进行比较。
- 用大写键入两行文字,其中包含并排放置在一起的问题字母,如 A、V 和 W 等。
- 调整字间距和行内字符间距,直至达到视觉平衡。

VARIOUS WAYS OF DRAWING WAVES

Original copy

WAV

WAV

WAV

1 Original copy
2 With kerning
3 With kerning and tracking

The Language and
Culture of Cartography
in the Renaissance

The Language and
Culture of Cartography
in the Renaissance

VARIOUS WAYS OF
DRAWING WAVES

Copy with kerning

VARIOUS WAYS OF
DRAWING WAVES

Copy with kerning and tracking

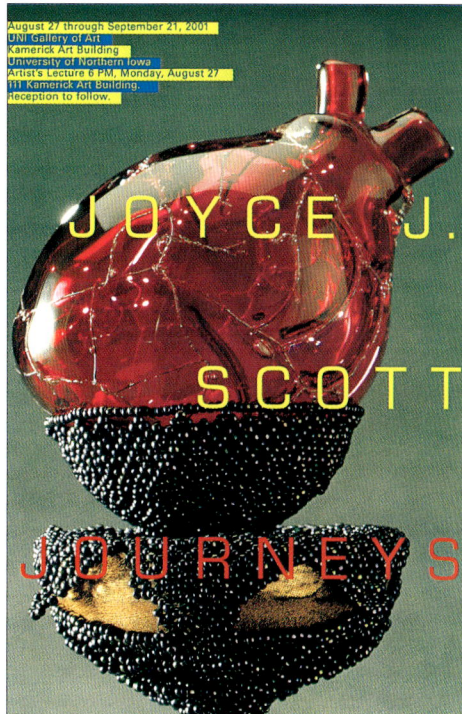

August 27 through September 21, 2001
UNI Gallery of Art
Kamerick Art Building
University of Northern Iowa
Artist's Lecture 6 PM, Monday, August 27
111 Kamerick Art Building.
Reception to follow.

JOYCE J.
SCOTT
JOURNEYS

1

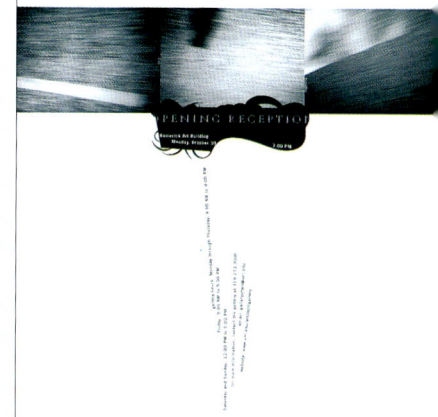

October 30 through
November 22, 200

University of Northern Iowa   Department of Art   Faculty Exhibition

OPENING RECEPTION

2

1. 在页面左上角使用的条块状底纹收到了良好的效果。当中较大的字体形成的开放空间与狭窄的条块状底纹形成了鲜明的对比。

2. 这张当代的设计作品聪明地使用版面设置达到了预期效果。注意看不合适的行间距造成的数字与字母的连接,以及字母没入页面上部和右边界的方式在增加视觉趣味性的同时也保证了可辨认性。

在版面设计中有两个基本的风格设置可以采用。一个称为对称风格，即设置围绕居中的中轴展开；这种风格通常被认为是传统的风格。另一种称为不对称风格，中轴不居中。这种风格被认为更具动感且能为设计增添更多的张力。

版面设置
风格、形状、格式、黑白翻转

页面设置

风格的设置也会影响到与之相应的图片的形状。但这相对来说比较简单，只要字块的形状不规则，居中或右对齐均可实现。

页面的格式也可能影响到字体的设置。风景格式的页面能够容纳宽度较大的字块，而肖像格式的页面字块相对较窄。

黑白翻转可以为设计增加戏剧性效果和冲击力。为了使翻转后的字迹清楚易读，最好采用无衬线的字体，因为纤细的衬线在印刷过程中可能会被覆盖。

设计效果： 1–3. 确保你选择的字体大小与页面格式匹配。在肖像格式页面上设置窄的字块，在风景格式页面上设置较宽的字块。居中（对称）方式可以使设计产生平和、传统的感觉，而不居中（不对称）的方式则使设计具有现代感的张力。4. 对称设置在使每一行文字都居中的同时也使其他的调节较为合理。5. 不对称的组合可以包含左对齐或右对齐。6. 尽量避免将两种风格混合在一起。7. 当采用黑白翻转风格时，最好用无衬线的字体，因为衬线字体的衬线往往容易被覆盖。8. 在黑白翻转的风格中，避免采用细衬线的字体，除非印刷质量比较高，否则字迹会比较模糊。9. 小于 8 磅的字体不要采用黑白翻转风格。

Landscape formats allow wider measures to be used.

Short measures get lost on a landscape format.

1

Short measures look better on a portrait format.

2

Do not use long measures on a portrait format.

3

Symmetry

Justified type aligns on both the left and the right. However, this style can look visually poor if the measure is too short, as the space between words can be excessive.

Symmetry

An interesting variation is to centre each line of type on the type measure. When used in conjunction with a centred heading, this gives a strong, symmetrical appearance.

use a
sans
if
you can

4

7

Range left

Ranged-left settings with an off-centred composition give an asymmetrical arrangement. This style tends to be more dynamic than a symmetrical style of setting.

Range right

As a variation, type can be aligned on the right of the measure. This has limited use because we are used to reading type from left to right; if too many lines are set in this style, the result can be hard to read.

hairlines
are
dangerous

5

8

Stick to your style

You should never mix styles in any element of design. This looks untidy, confused and disjointed, and is difficult to follow.

Style

Make your decision before you start work and then stick to it. You will find it far easier to achieve a stylish, clear design.

beware
of
being too small

6

9

练习十九

- 选择一张 6 × 9 英寸的页面，用无衬线字体的粗体不分大小写键入如下文字："keep off the grass"。
- 按照你认为最好的组合方式设置页面。
- 用 Bodoni 字体重复上述练习。
- 比较哪种字体的效果更好。
- 选择包含 50 到 60 个单词的一小块文字。
- 分别用 10 磅的 Bodoni、Century Schoolbook 和 Bembo 字体设置文本。
- 将每一种字体的文本黑白翻转。
- 打印设计出的页面，看看字体是否容易辨认。
- 使用无衬线字体的细体、正常体和粗体重复上述练习。
- 将同样的一段文字用各种有趣的形状进行排列，并加上一张图片，看看对文字的可读性和页面设计效果的影响。

**Keep off
the Grass**

**Keep off
the Grass**

How it all started

The basic structure of the typefaces we use today was established by calligraphers at the end of the 15th century. They took their inspiration on the one hand from Roman capitals and on the other from the manuscript styles known as Carolingian minuscules, which were established in the reign of the Emperor Charlemagne in the latter part of the 8th century.

练习二十

- 选择一张 6 × 9 英寸的风景或肖像风格的页面，用衬线字体键入如下文字："Symmetry: The traditional typographic method of layout whereby lines of type are centered on the central axis of the page。Balance is achieved by equal forces"。
- 用对称风格进行排版。
- 用无衬线字体设置如下文字："Asymmetry: The dynamic method of layout used by modernists whereby lines of type are arranged on a non-central axis。Balance is achieved by opposing forces"。
- 用不对称风格进行排版。
- 比较两种风格的不同之处。

Symmetry

The traditional typographic method of layout whereby lines of type are centred on the central axis of the page. Balance is achieved by equal forces.

Asymmetry

The dynamic method of layout used by modernists whereby lines of type are arranged on a non-central axis. Balance is achieved by opposing forces.

How it all started

The basic structure of the typefaces we use today was established by calligraphers at the end of the 15th century. They took their inspiration on the one hand from Roman capitals and on the other from the manuscript styles known as Carolingian minuscules, which were established in the reign of the Emperor Charlemagne in the latter part of the 8th century.

How it all started

The basic structure of the typefaces we use today was established by calligraphers at the end of the 15th century. They took their inspiration on the one hand from Roman capitals and on the other from the manuscript styles known as Carolingian minuscules, which were established in the reign of the Emperor Charlemagne in the latter part of the 8th century.

How it all started

The basic structure of the typefaces we use today was established by calligraphers at the end of the 15th century. They took their inspiration on the one hand from Roman capitals and on the other from the manuscript styles known as Carolingian minuscules, which were established in the reign of the Emperor Charlemagne in the latter part of the 8th century.

It came from
outer space

Irregular shapes are easy to achieve with computer setting. They will add visual interest to a piece of text and at the same time they can reflect the contents.

It came from
outer space

Irregular shapes are easy to achieve with computer setting. They will add visual interest to a piece of text and at the same time they can reflect the contents.

It came from
OUTER SPACE

Irregular shapes are easy to achieve with computer setting. They will add visual interest to a piece of text and can reflect the contents.

线条和修饰可以指导读者阅读，或吸引读者对页面中某部分的
注意。它们同样也可以作为页面的装饰物。

版面设置
线条和修饰

线条和修饰可以在正文和标题中使用

　　各种粗细和风格的线条唾手可得（如直线、虚线、破折号等）。线条的粗细从半磅（极细）开始。当线条用在正文中时，可以帮助读者区分正文内容和小标题。垂直的线条可以将正文划分成数块；水平的线条则是组织信息、帮助阅读的良好工具。同时，线条也可以用在单词下方，起到对正文某些内容的强调作用。

　　在标题设置中，线条可以造成戏剧化的效果，引起读者的注意。如果线条足够宽，可以采用黑白翻转的方式。在广告招贴设计中，线条甚至可以成为重要信息的标识符。

　　修饰（也可称为花边）经常被印刷商和设计师用来装饰页面。这种风格发源于中世纪的书籍中，那些书籍的页面装饰繁杂，页边也饰以印刷的装饰线。最常用的装饰页面的方法是插入 zapf 印刷符号或类似的修饰物。

设计效果：1. 在表格式的页面中，线条用以区分信息。2. 此处线条用以帮助识别项目。3. 在目录页中，线条可以派上大用处。4. 在标题设置中，可以用对比的方式为标题加上下划线，起到强调的作用。例如，36磅 univers light 字体的标题配上较黑的下划线，而较粗的字体则配以较细的下划线。5. 线条也可用于引起读者对某一部分内容的重视和注意。6. 修饰造成的"传统"效果。

Title	Name	
Address		
		Postcode
Email		
Tel	Fax	

1

Series D				
The Fairy Queen	10	September	20.30h	2
Ariadne auf Naxos	23	October	20.30h	19
Recital Gosta Winbergh	7	January	20.30h	53
Norma	14	January	20.30h	59
Pikovia Dama	7	February	20.30h	67
Il viaggio a Reims	12	March	20.30h	78
Orfeo ed Euridice	16	April	20.30h	96

2

Contents

3

In magazine work, typographers must constantly look for ways of attracting attention and stimulating the reader.

<u>Rules draw attention to subheads</u>

The way subheadings and headings are positioned is one method of bringing the reader into the text. Rules can give dramatic tension to a heading.

5

Univers
Light

Univers
Black

4

BRITISH PAINTING

❧

1948–1964

❧

6

练习二十一

- 选择一张6×9英寸的页面，为任意一本书的目录页排版。字体大小、字块宽度和行间距按照你的意愿设置。页码右对齐。
- 采用同样的设置，在每一个项目的下面加上细线或1磅的线条。
- 比较这两种设计的效果。

Contents

Contents

练习二十二

- 使用与练习二十一中相同的页面格式，将如下文字用大写或区分大小写重复设置三遍："Rules have Power"。
- 首先将字体设置为无衬线字体的36磅细体，并加上12磅的下划线。
- 接着将字体设置为无衬线字体的36磅正常体，并加上6磅的下划线。
- 将字体设置为无衬线字体的36磅粗体或特粗体，并加上1磅的下划线。
- 分析各种设置的效果。

Rules have Power

Rules have Power

Rules have Power

Rules have Power

Rules have Power

Rules have Power

Rules have Power

Rules have Power

Rules have Power

线条和修饰

We *make our* CUSTOMERS *and keep their* COMPUTERS *and* **productive** safe. reliable. Anywhere. Anytime.

Getting more
from what you've
already got.

AT SYMANTEC, WE'RE DEDICATED to the notion that computing should be simpler, faster, and more productive. That's why we build products that automatically resolve problems, or make things easier to do.

Take our *Norton Utilities*™ software, for example. If the information on your PC is mission critical—and what business information isn't?—*Norton Utilities* gives you peace of mind by protecting the integrity of your data and keeps your system running at peak performance. With *Norton Utilities*, you can always count on the comfort of knowing that if it isn't broken, we'll fix it anyway.

Our *Norton AntiVirus*™ products operate under the same preventive-medicine principle: it is better to not have a virus than to recover from one. Which is why we've set up the Symantec AntiVirus Research Centers (SARC™)—world-renowned research institutes devoted to locating viruses and rooting them out. So you can get on with your work without having to worry.

How can a contact manager make you more productive? By doing a lot more than just holding names and addresses. Our *ACT!*™ software gives you all the information you need to stay on top of managing every aspect of all your business relationships. And if you're working away from the home office, products such as *pcANYWHERE*™ and *WinFax PRO*™ will make things so familiar as to make you think you're right at your own desk.

What about the Internet? With technologies changing so rapidly, your developers need a set of tools that will allow them to get their web development applications to market before they become obsolete. This is precisely the idea behind our award-winning *Symantec Café*™ series of authoring products. With Symantec Café, it's extremely easy for both novice and experienced programmers alike to develop dynamic Java™ applications and applets.

And because business these days is more global than ever, it's important that people anywhere feel comfortable with our products. So we adapt our software to local markets and cultures—18 languages and counting.

Add it all up and you'll find that you have a multitude of ways to make life on your PC easier, better, and more rewarding.

Who could ask for more?

1

NOTE NUMBER 169

DECEMBER 1999

THE WORLD BANK GROUP · FINANCE, PRIVATE SECTOR, AND INFRASTRUCTURE NETWORK

viewpoint

Gas Reform in Ukraine

Monopolies Markets Corruption

Private Participation in theReform of the natural gas industry in Ukraine started a year later than reform of the power industry. Because gas reform had no blueprint, its direction has remained ambiguous. Private Participation in theReform of the natural gas industry in Ukraine started a year later than reform of the power industry. Because gas reform had no blueprint, its direction has remained ambiguous. PrivatParticipation in theReform of the natural gas industry in Ukraine started a year later than reform of the power industry. Because gas reform had no blueprint, its direction has remained ambiguous.

Reform of the natural gas industry in Ukraine started a year later than reform of the power industry. Because gas reform had no blueprint, its direction has remained ambiguous. Reform of the natural gas industry in Ukraine started a year later than reform of the power industry. Because gas reform had no blueprint, its direction has remained ambiguous. Reform of the natural gas industry in Ukraine started a year later than reform of the power industry. Because gas reform had no blueprint, its direction has remained ambiguous. Reform of the natural gas industry in Ukraine started a year later than reform of the power industry. Because gas reform had no blueprint, its has remained ambiguous.

Private Participation in theReform of the natural gas industry in Ukraine started a year later than reform of the power industry. Because gas reform had no blueprint, its direction has remained ambiguous. Private Participation in theReform of the natural gas industry in Ukraine started a year later than reform of the power industry. Because gas reform had no blueprint, its direction has remained ambiguous. PrivatParticipation in theReform of the natural gas industry in Ukraine started a year later than reform of the power industry. Because gas reform had no blueprint, its direction has remained

natural gas industry in Ukraine started a year later than reform of the power industry. Because gas reform had no blueprint, its direction has remained ambiguous. Private Participation in theReform of the natural gas industry in Ukraine started a year later than reform of the power industry. Because gas reform had no blueprint, its direction has remained ambiguous. PrivatParticipation in theReform of the natural gas industry in Ukraine started a year later than reform of the power industry. Because gas reform had no blueprint, its direction has remained ambiguous.

2

---

1. 此处是线条与大字体和小字体的组合使用范例。对标题和页眉进行的强调形成了具有时尚感的装饰效果。标题中特别有趣的地方是线条穿过了低于 x 坐标的字母的下部分。
2. 标题上下方的线条将注意力吸引了过来。标题下方线条中的文字则显示出了线条的另一种妙用。垂直的线条将正文划分开来，形成了清楚的格式，最大限度地容纳了可以阅读的文字。

色彩是设计师进行设计时最活跃的元素。它不仅为设计增添了变化和情趣，还增加了设计的空间感。今天的设计师们是幸运的：相关软件的开发使色彩的操纵和运用比过去容易得多。

色彩
信息

选择颜色

如同字体能向我们传达出信息一样，色彩给我们的信息更多。记住色彩具有的象征意义是非常重要的。例如红色，往往让人联想起火焰，因而使人觉得温暖并充满力量。而蓝色则使人感受到宁静和淡漠。你选择的颜色会影响作品的情趣和人们的回应程度。

特殊的色彩组合也可以造就设计的情趣。要产生和谐的感觉，就使用相近的颜色，即色谱中邻近的两种颜色，例如蓝色和绿色。要具有更多的张力和变化，就使用对比色，即色谱中相对的两种颜色，如红色和绿色。这种组合虽然会产生冲突，但可以迅速引起注意。

设计效果： 1-2. 某些颜色使人信心百倍，而某些颜色却让人觉得心灰意冷。因此如果你要设计出催人奋进的作品，红色和橙色是设计中最好的选择。3-4. 当紧随红色之后时，蓝色和绿色往往不那么激进。5. 相互协调的颜色在色谱中的距离较近，如黄色和绿色，或者蓝色和绿色。互补色指的是色谱中相对的颜色，如红色和绿色，或黄色和紫色。这样的组合充满了变化和张力。6. 黄色比较柔和，不如红色醒目。7. 同一色系的颜色也有差别（见58页）。通过细小的调整，你可以在同一色系内形成各种变化。

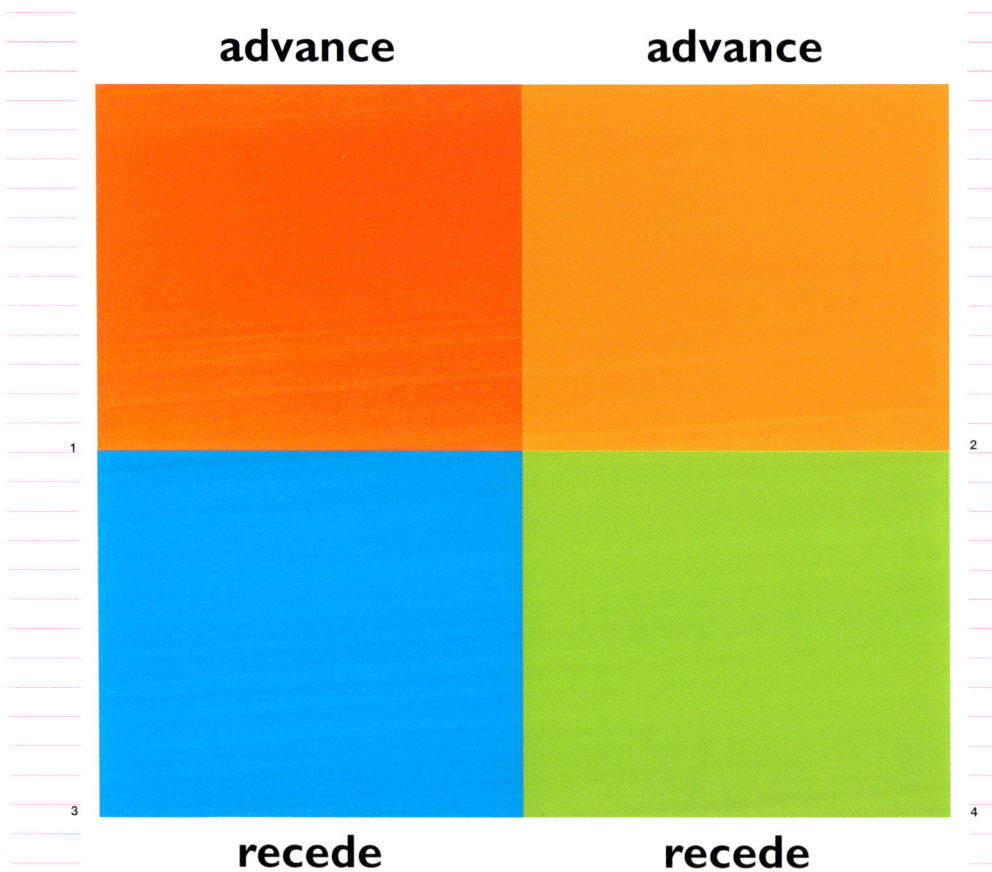

advance   advance

1   2

3   4

recede   recede

5

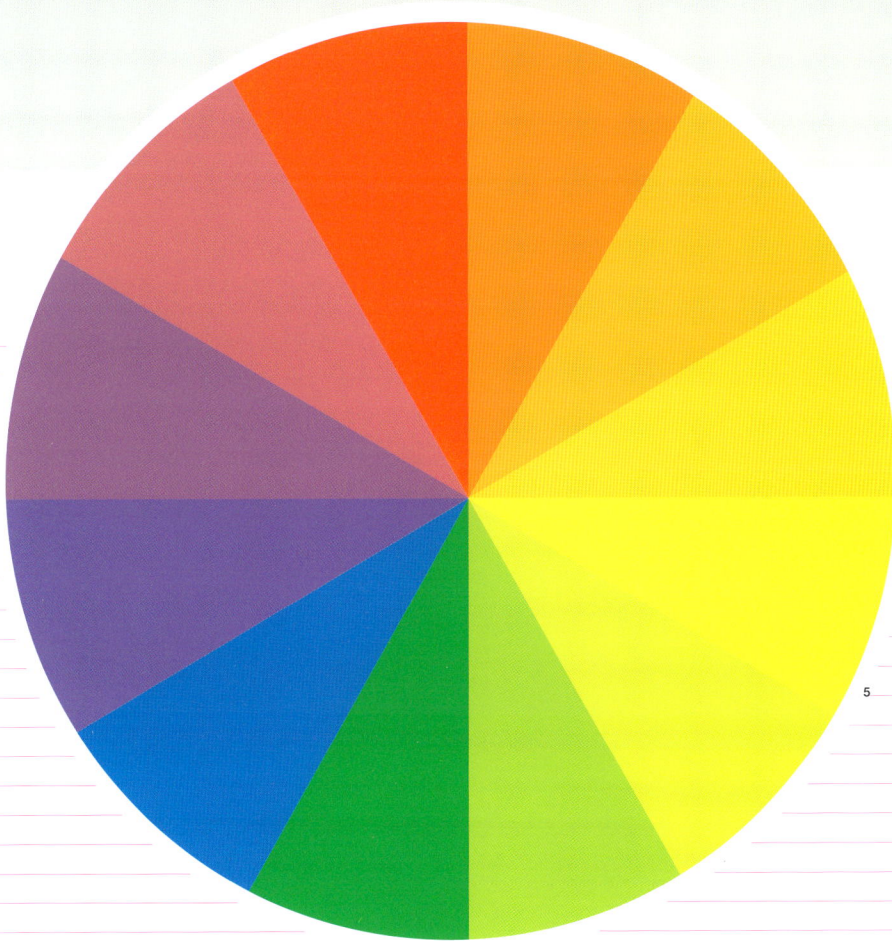

There are many intensities

rock'n'roll

rock'n'roll

rock'n'roll

rock'n'roll

rock'n'roll

Do these colours advance or recede?

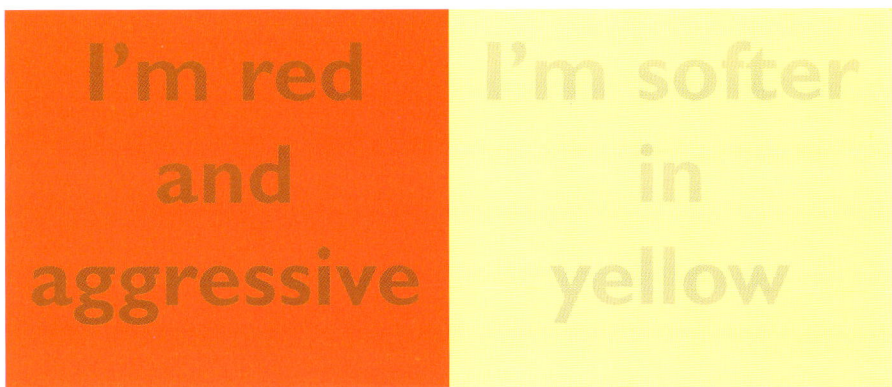

I'm red and aggressive

I'm softer in yellow

6

7

练习二十三

- 选择一张4×6英寸的页面，用无衬线字体的粗体键入大写的如下文字："BRAVO！"，并设置成大红色。
- 将字体设置为宝石蓝色。
- 将两者并排放置，并比较其效果。
- 用相同的页面和字体设置如下文字："JAZZ BAR"。也分别设置为红色和宝石蓝色，但尝试在同一色系内对颜色进行改变。

BRAVO!　　　**BRAVO!**

JAZZ BAR　　　JAZZ BAR

练习二十四

- 使用与练习二十三中相同的格式、文字、字体和颜色，将红字的背景分别设置为黑色和黄色。
- 自己选择文字重复上述练习，看看不同的背景对不同信息所产生的影响。
- 打印你的设计并比较效果。

BRAVO!　　　**BRAVO!**

JAZZ BAR　　　**JAZZ BAR**

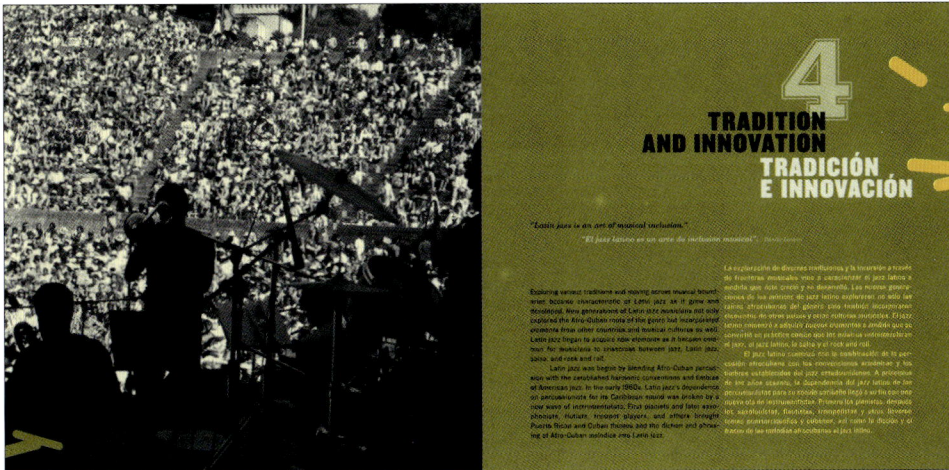

1. 颜色可以区分标题和正文中不同的内容。在这里，色彩被成功地用来区分两种不同的语言。对纯色进行翻转相对来讲比较容易，同时也可以为设计增添魅力。
2. 这个范例展示了对色彩非常成功的运用。单词 "inside" 嵌入背景中，很好地体现了文字本身的含义。
3. 红、蓝两种原色的选用显示了色彩对设计的强化作用。此外，色彩还可以诠释文本的含义。

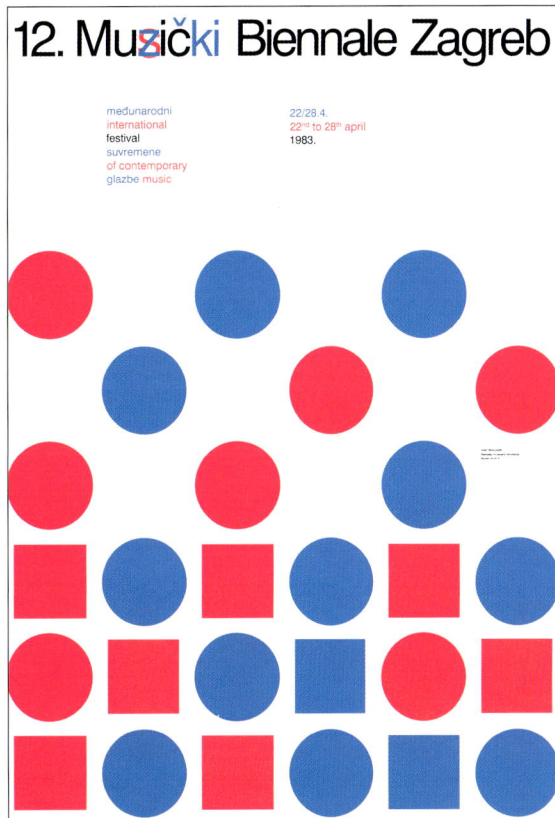

面对色彩的使用可能爆发出的无穷变化，有一条规则需要遵守：不要过度使用色彩，否则你将失去它的效果。

色彩
区别与易读性

关于颜色的术语

在处理色彩时必须理解关于颜色的三个术语：基色——纯粹的颜色，如红色、蓝色和绿色；色调——同一种颜色的深浅变化；纯度——色彩浓烈程度的范围，如高浓度的大红色和低浓度的深绿色。

当颜色和字体组合在一起时，文字的易读性要靠颜色的对比来保证。对比最大的是白底黑字，对比最小的是白底黄字。这两者之间有很大的一个可以变化的范围。背景颜色与字体颜色越接近，文字的易读性就越低。

设计效果：1、色彩被按照三种方式加以区分。2、基色。各种颜色的固有名称。3、色调，颜色由亮到暗的变化。4、纯度，色彩由高纯度到低纯度的变化范围。5-6、最强烈的对比是白底黑字或黑底黄字。7、最弱的对比是白底黄字。8-10、随着背景和字体颜色的接近，文字的易读性逐渐降低。11、字体风格也会影响颜色的表现力。有衬线的字体可供上色的面积不大，因此推荐采用纯度较低的颜色，使字体看上去轻盈一些。12-13、无衬线的字体或黑体能够很好地体现色彩的感觉，因此颜色的纯度要高。

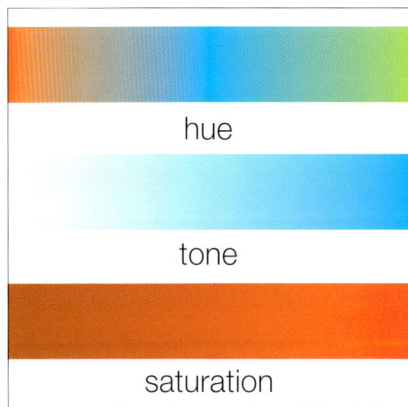

hue

tone

saturation

1

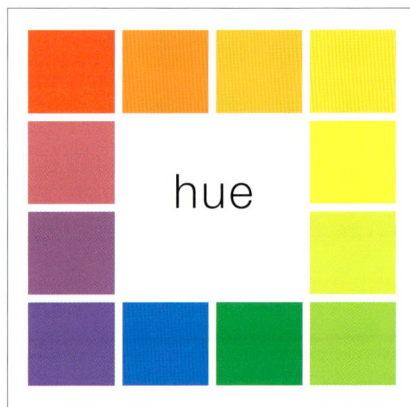

hue

2

tone

3

saturation

4

Taxi

Taxi

Taxi

Taxi

Taxi

Taxi

Taxi

Taxi

Taxi

5

6

7

8

9

10

11

12

13

练习二十五

- 选择一张8.5×11英寸的风景格式的页面。
- 用无衬线字体的粗体键入"OK！"，并设置为白底黄字。
- 重复上述练习，将字体颜色依次设置为橙色、红色、绿色、蓝色和紫色。
- 分析得出的结果，并指出哪种颜色给人的感觉较强烈，哪种较弱。
- 调整背景和字体的颜色，尝试对比最大和对比最小的搭配。

练习二十六

- 使用与练习二十五中相同的格式，用12磅Bembo字体键入"Bar Manager"，并设置为白底红字。
- 重复上述练习，将字体改为12磅无衬线字体的粗体。
- 观察两者的效果，看看哪一种的色彩纯度较高。

1. 高纯度的色彩使画面非常和谐，同时也增强了图片的感染力。
2. 使用颜色增强或弱化图片的范例。仅仅改变了设计的背景颜色，整个设计的实际效果就发生了截然不同的变化。
3. 对颜色的渲染可以使画面柔和且富有生气。此处对背景中进行的橙色渲染增加了页面的空间感。
4. 两种纯度非常高的色彩可以营造出极具冲击力的效果。

1

2

3

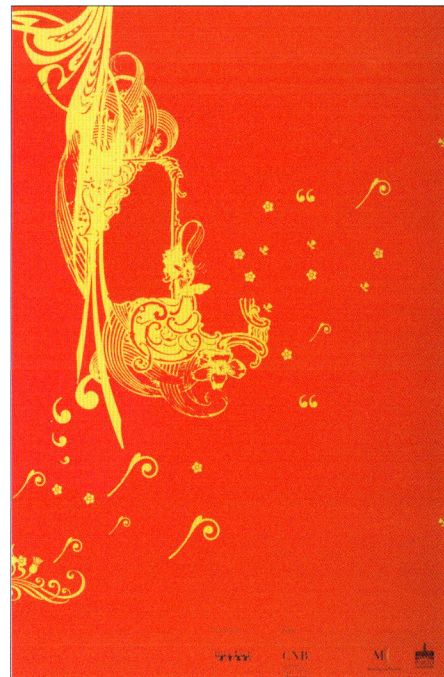

4

伴随着大量精细软件和印刷机的问世，将字体、颜色和图片进行组合，既迅速又便捷，同时产生了许多意想不到的效果。

色彩
字体、颜色和图片的组合

颜色的冲击力

随着五到六色印刷机的问世，设计师可以将四种颜色（四分色）和其他的一至两种颜色结合，达到色彩的最佳效果。

四分色众所周知，是将青、品红、黄色和黑色组合在一起的减色系统。这种系统最早使用于印刷机。四种颜色的组合可以产生其他的颜色。

调和色是为设计某种独特的颜色而预先设置好的。这些颜色有相关的标准，从而使得印刷机和设计师能够通过数字调节来使颜色保持一致。

这种系统使设计师能充分运用各种颜色创造出与小样或照片中一样的色彩效果。

颜色可以用来作为背景衬托出字体，或者可以融入整个页面中。就好像你先搭出一个盒子，再往里面添加色调较淡、纯度较低的颜料（例如 20% 的黄色），因此而造就了一个良好的背景，黑字在上面显得异常清晰。对色彩进行的分级可以被充分运用于创造丰富的背景。同样，不同的颜色或相同颜色的不同级别可以用来突出正文部分的某些内容。而为了保证文字的易读性，通常只有较大的标题才采用色彩渲染。

1

设计效果：1. 在大的标题字体中，当字母重叠在一起时，色彩可以造就出新的有趣的形状。2. 两种颜色叠加在一起会产生第三种颜色，为设计增加视觉效果。3. 白色的文字和深蓝色的天空形成强烈的对比，使文字容易读取。这种翻转式的风格也是一种在图片上合理利用空白空间的较为经济的方式。4. 背景可以着上精巧的颜色，为文字的排列造就良好的载体。5. 嵌入彩色的边框是另一种增加页面变化的方式。6-8. 颜色的分级也增加了视觉表现力。

Rock n' Roll

2

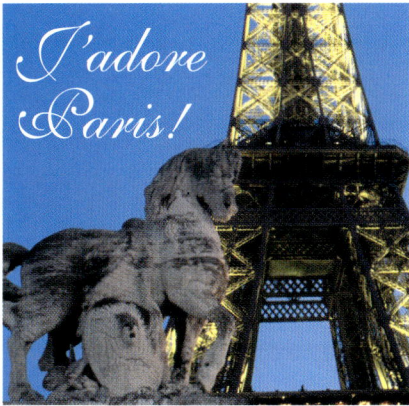

J'adore Paris!

3

The creativity of colour

Colour attracts the reader: it opens up the whole design and gives an added dimension to the visual look of a piece. By changing the density of the colour you are using, you can make it look as if another colour is being used — and, of course, it creates contrast within the overall composition.

4

The creativity of colour

Colour attracts the reader: it opens up the whole design and gives an added dimension to the visual look of a piece. By changing the density of the colour you are using, you can make it look as if another colour is being used — and, of course, it creates contrast within the overall composition.

5

The creativity of colour

Colour attracts the reader: it opens up the whole design and gives an added dimension to the visual look of a piece. By changing the density of the colour you are using, you can make it look as if another colour is being used — and, of course, it creates contrast within the overall composition.

6

The creativity of colour

Colour attracts the reader: it opens up the whole design and gives an added dimension to the visual look of a piece. By changing the density of the colour you are using, you can make it look as if another colour is being used — and, of course, it creates contrast within the overall composition.

7

The creativity of colour

Colour attracts the reader: it opens up the whole design and gives an added dimension to the visual look of a piece. By changing the density of the colour you are using, you can make it look as if another colour is being used — and, of course, it creates contrast within the overall composition.

8

练习二十七

- 选择一张6×9英寸的风景格式页面。
- 键入几个数字,并调整字体至充满页面的中间位置。
- 选择两种原色。
- 将不同颜色的数字叠加,使叠加部分构造出新的形状。
- 两种原色叠加的地方产生了新的颜色。

练习二十八

- 采用与练习二十七中相同的格式,构造一个3平方英寸的正方框,并填进20%的黄色。
- 用10磅的无衬线字体键入30个单词。
- 在20%的黄色背景中将单词设置为黑色。
- 重复上述练习,在正方框的边缘加上2磅的线条并填入60%的黄色。
- 尝试逐渐增加黄色的纯度至饱和。
- 观察你的试验。
- 使用同样的正方框,使背景由一种颜色从上到下渐变。
- 尝试色彩与背景反字体的组合,以及色彩与渐变方向的变化。

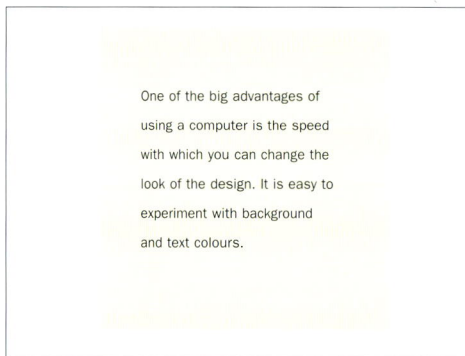

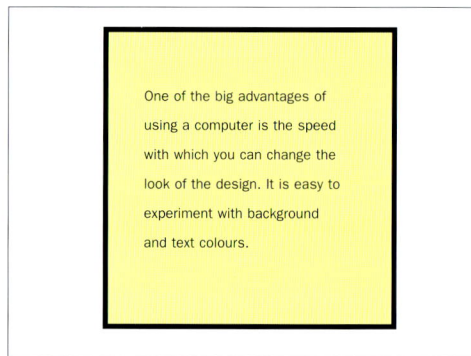

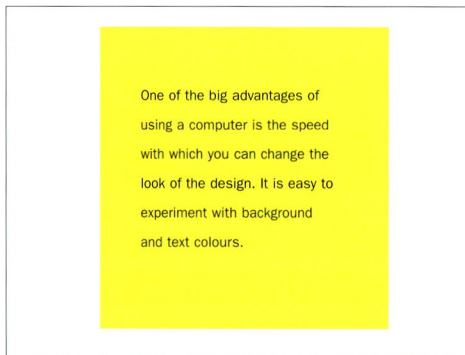

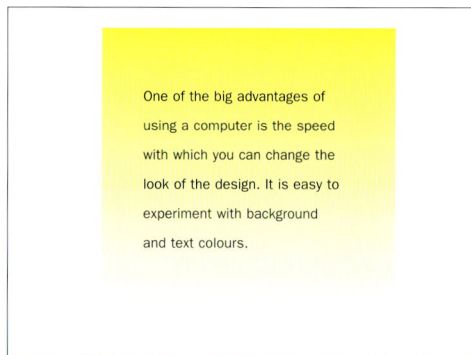

One of the big advantages of using a computer is the speed with which you can change the look of the design. It is easy to experiment with background and text colours.

1. 在这个引人注目的范例中，色彩帮助读者区分不同的项目。同样也增强了我们对图片和文字的视觉感受。

2. 色彩是强调某个项目的好方法。图片中和文字中使用的明亮颜色说明了这一点。

3. 灰色通常作为中间色平衡图片和文字。中间色的使用可以减少由于色彩浓重或过分张扬造成的冲突。

4. 图片中成功地使用了柔和的色彩与黑暗的人物背影形成对比，并达到了和谐。

在决定选择何种字体之前，你需要考虑正文中的优先性。也即是说要建立文字的等级（强调的轻重顺序），这样你才能知道你在处理的文字中有多少个不同的层次。

信息解析
正文的等级

重要性的不同层次

对于标题，你可能会遇到三种层次，你可以命名为 A、B 和 C。正文部分也需要按照某种方式加以区分，以帮助你建立起每一层次信息的标准设置。

不同层次的标题可以通过字体和磅值的大小加以区分。对正文而言，字体大小只是一种区分方式。除此之外，改变字体风格（如变无衬线字体为有衬线字体）也是一种方法。最后，你还可以使用色彩对此加以区分。

设计效果：1. 这张音乐节目单按照重要性分有五个层次：节目号、日期、时间、作曲者、音乐片断及演奏者。2. 在设置页面格式时，字体（如将字体变为斜体）和色彩的对比都是区分不同层次信息的好方法。3. 或者也可以在字体风格上形成对比。4. 当标题和正文同时出现时，不同大小和风格的字体可以灵活运用。5. 线条也可以在区分信息层次时产生积极的作用。6. 方框用以强调某部分正文。

---

37 Saturday August 17
7.30pm–9.50pm

Rodgers, orch. Hans Spialek Babes in Arms –
Overture **Rodgers, orch. Robert Russell
Bennett** Victory at Sea – Symphonic Scenario;
On Your Toes – Slaughter on Tenth Avenue
Rodgers & Hammerstein Oklahoma!
(concert version)
Maureen Lipman, Lisa Vroman
Klea Blackhurst, Tim Flavin, Brent Barrett

1

Wednesday

● **Susan Bullock**
Soprano joins the South
Quarter Sinfonia for works
by Ravel and Wagner.
St Mark's Chapel (07 22 1061)
7.30pm £8–£15.

● **Ian Fountain**
Piano works by Beethoven.
*St James's church
(02 38 0441)
7.30pm £7–£13.*

2

36
FAMILY FAVOURITES
Cornish Ware brought to book.

38
LATE QUARTET
Linen Works: a group of weavers with more
than one string to their bow.

42
TIME FOR CRAFTS TO
GO TO THE BALL
Arts for Everyone – will the new lottery
scheme live up to its name?

44
SOURCES OF INSPIRATION
The sculptor Bryan Illsley talks about his
life and work.

3

House Specialties

SERVED WITH PILAU RICE

Chicken Passanda.........£6.50

A mild and delightful dish, specially cut slices of chicken marinated with yoghurt based sauce and cooked in fresh cream, mixed ground nuts and almond powder.

Karhai Lamb...............£6.50

Tender pieces of lamb grilled in a tandoori oven, cooked with garlic, ginger, tomatoes, onions, capsicum and fresh coriander, medium spiced.

Karhai King Prawn.......£8.95

Achar Gost.................£6.50

A fairly hot dish. Pieces of marinated lamb cooked in a tantalizing pickle masala, laced with whole green chillies.

Amere Murgh..............£6.50

A delightful mild chicken dish, cooked with pulp mango, mild spices, fresh cream and almonds.

4

Best Saving Rates

Find the best deals on savings accounts

PROVIDER	PRODUCT	GROSS RATE
£1/£10 to invest		
Rock	Online Saver	3.35
Net	Online Saver	3.05
Local Friendly	Immediate Access	3.00
£1,000 to invest		
Southern Reliant	Saver Plus	3.40
Roots	Online Saver	3.35
BankDirect	Saver Plus	3.22
£25,000 to invest		
BankDirect	Midi Longterm	3.40
Roots	Saver Plus	3.35
Local Friendly	Midi	3.25
£50,000 to invest		
Roots	30 Day Access	4.20
Net	3 Year Access	3.75
Local Friendly	2 Year Access	3.55
Southern Reliant	Midi Mini	3.45
Lane and Lane	Midi Saver	3.40
Over £50,000 to invest		
Lane and Lane	Midi Maxi Longterm	5.55
Northern Circle	Fixed Saversure	5.20
Southern Reliant	Fixed Saversure	4.75
Local Friendly	Saversure Access (variable)	4.70
Net	Premier Plus	4.65

5

DVD & VIDEO

The Magnificent Seven
Stylish

★★★

Starring: Steve McQueen, James Coburn, Charles Bronson, Robert Vaughn and Yul Brynner. The beleaguered denizens of a Mexican village, weary of attacks by banditos, hire seven gunslingers to repel the invaders. This is without a doubt a film to remember.

Gone with the Wind
Old classic – not to be missed

★★★

More than a movie: this 1939 epic (and all-time box-office champ) is superbly emotional and nostalgic. Vivien Leigh is magnificent; Clark Gable provides one of the most charismatic performances ever. It's an achievement that pushed its every resource – art direction, colour, sound, cinematography – to new limits.

A Beautiful Mind
High interest

★★★★★

A decent biopic of the Nobel prize-winning mathematician, John Nash. It doesn't come close to conveying the contribution Nash made to economics, and it's really just Shine with sums, repeating as it does one of Hollywood's favourite equations – genius equals madness.

6

练习二十九

- 选择一张 6 × 9 英寸的肖像格式页面，用 9 磅或 11 磅字体键入一段 200 字左右的文章。
- 分析这段文字，并为之加上大标题、副标题和你认为相关的图片。
- 采用同样的格式，利用字体大小、风格和磅值的变化区分出如下八个层次的信息：1；Friday July 19；7：30pm；Haydn；The Creation；featuring；Christiane Delze；Soprano。
- 调整字体、颜色和风格。

Each year the second year students, in conjunction with their Contextual Studies program, produce a course magazine. The subject matter relates to the design industry. Each student is required to write approximately 500–750 words on a subject of choice, research the appropriate visual material, and bring these elements together in a design for a double-page spread.

The objective of this project is to enable students to oversee a complete concept from the selection of the subject matter and writing of an article, to the design and production of the magazine.

Students are encouraged to: use their writing skills; analyse subject matter; make constructive critical comments.

The subject matter for the article is of the student's own choosing, thus motivating them to choose a subject of interest. This in turn encourages care and clarity in the treatment of the text and the use of language.

The work is assessed according to the following guidelines: the student's ability to evaluate in written form a design topic; the application of appropriate visual material to improve the article and provide an effective design; ability to meet production deadlines.

Magazine Design Project

Each year the second year students, in conjunction with their Contextual Studies program, produce a course magazine. The subject matter relates to the design industry. Each student is required to write approximately 500–750 words on a subject of choice, research the appropriate visual material, and bring these elements together in a design for a double-page spread.

Aims and Objectives:

The objective of this project is to enable students to oversee a complete concept from the selection of the subject matter and writing of an article, to the design and production of the magazine.

Students are encouraged to:
- use their writing skills
- analyse subject matter
- make constructive critical comments.

The subject matter for the article is of the student's own choosing, thus motivating them to choose a subject of interest. This in turn encourages care and clarity in the treatment of the text and the use of language.

Assessment criteria:

The work is assessed according to the following guidelines:
1. The student's ability to evaluate in written form a design topic.
2. The application of appropriate visual material to improve the article and provide an effective design.
3. Ability to meet production deadlines.

练习三十

- 选择一段 200 字左右的文章，并在 6 × 9 英寸的肖像格式页面中将之设置为 10 磅 Times 字体。
- 分析这段文字，并为之加上大标题、小标题，区分出信息的不同层次。
- 在 8.5 × 11 英寸的页面中构建一个包含两列的网格，并将正文和标题按照一定的风格进行排版。
- 在同样尺寸的页面中构建一个包含两列的网格。排入标题和正文，并加入两张图片，设置页面的风格。

Graffiti is a contentious subject. It can be viewed either as vandalism or as an art form. On the one hand people argue that it damages our environment and should not be condoned; on the other hand it is seen to have an intense energy and creativity. Whatever your opinion, you cannot ignore the fact that graffiti is now part of urban life. The essence of the best graffiti can be utilised as a starting point for design concepts. Speed, vigor, and excitement are inseparable from graffiti art. This is because it is usually an illicit activity. Its clandestine nature is in itself a motivating force and dictates the type of equipment used: spray cans or broad-tipped markers – easily carried, effective, and fast. The rich visual dynamism of designed mark-making conveys a great sense of enjoyment that may well be echoed in the viewer: Graffiti has many lessons to offer in color and pattern. There are uses and applications for this questionable street art. Urban locations have been designated for this very purpose. Record covers, posters, comics, and badges use this medium to capture the essence of a message and convey it in an interesting and powerful way.

Graffiti

Graffiti is a contentious subject. It can be viewed either as vandalism or as an art form. On the one hand people argue that it damages our environment and should not be condoned; on the other hand it is seen to have an intense energy and creativity.

Not to be ignored

Whatever your opinion, you cannot ignore the fact that graffiti is now part of urban life. The essence of the best graffiti can be utilised as a starting point for design concepts. Speed, vigor, and excitement are inseparable from graffiti art. This is because it is usually an illicit activity. Its clandestine nature is in itself a motivating force and dictates the type of equipment used: spray cans or broad-tipped markers – easily carried, effective, and fast.

Street art

The rich visual dynamism of designed mark-making conveys a great sense of enjoyment that may well be echoed in the viewer: Graffiti has many lessons to offer in color and pattern. There are uses and applications for this questionable street art. Urban locations have been designated for this very purpose.

Record covers, posters, comics, and badges use this medium to capture the essence of a message and convey it in an interesting and powerful way.

1

Friday July 19 7.30 pm

Haydn

The Creation

FEATURING

Christine Delze

Soprano

1

HAYDN

The Creation

FEATURING

Christine Delze

Soprano

FRIDAY JULY 19
7.30 PM

Graffiti

Graffiti is a contentious subject. It can be viewed either as vandalism or as an art form. On the one hand people argue that it damages our environment and should not be condoned; on the other hand it is seen to have an intense energy and creativity.

Not to be ignored
Whatever your opinion, you cannot ignore the fact that graffiti is now part of urban life. The essence of the best graffiti can be utilised as a starting point for design concepts. Speed, vigor, and excitement are inseparable from graffiti art. This is because it is usually an illicit activity. Its clandestine nature is in itself a motivating force and dictates the type of equipment used: spray cans or broad-tipped markers – easily carried, effective, and fast.

Street art
The rich visual dynamism of designed mark-making conveys a great sense of enjoyment that may well be echoed in the viewer: Graffiti has many lessons to offer in color and pattern. There are uses and applications for this questionable street art. Urban locations have been designated for this very purpose.

Record covers, posters, comics, and badges use this medium to capture the essence of a message and convey it in an interesting and powerful way.

Graffiti

Graffiti is a contentious subject. It can be viewed either as vandalism or as an art form. On the one hand people argue that it damages our environment and should not be condoned; on the other hand it is seen to have an intense energy and creativity.

Not to be ignored
Whatever your opinion, you cannot ignore the fact that graffiti is now part of urban life. The essence of the best graffiti can be utilised as a starting point for design concepts. Speed, vigor, and excitement are inseparable from graffiti art. This is because it is usually an illicit activity. Its clandestine nature is in itself a motivating force and dictates the type of equipment used: spray cans or broad-tipped markers – easily carried, effective, and fast.

Street art
The rich visual dynamism of designed mark-making conveys a great sense of enjoyment that may well be echoed in the viewer: Graffiti has many lessons to offer in color and pattern. There are uses and applications for this questionable street art. Urban locations have been designated for this very purpose.

Record covers, posters, comics, and badges use this medium to capture the essence of a message and convey it in an interesting and powerful way.

在某些方面，插图或图表也许会比图片更具动感和吸引力。插图能直观地表现某个内容，强调某个方面。

信息解析
插图和图表

插图的指引作用

在对正文信息进行分析时，插图的使用可以令复杂的问题变得清晰明了。因此插图也被用于对信息的传达进行辅助。这里有许多使用插图的理由：它们可以对读者起到启发作用，可以体现出问题是如何被解决的，可以为读者指明目标，可以跨越语言的障碍，形成情感呼应，甚至可以诠释从未存在过的情形。

作为一个设计师，你可以利用计算机绘制地图和图表、生成图形。关键在于如何才能绘制出合适的图表。首先要确保图表中的任何文字均能被辨认出来，并且图表的引入简化了问题，而不是使之复杂化。记住一句古老的格言："简约胜过繁复！"

设计效果：1. 地图是图表中常常被采用的一种比较有效的形式。这张地图通过三维图像使之更清晰易懂。2. 插图被用来对产品进行说明。照相机是精密的器械，用简单的图表对各部分进行标注，可以帮助使用者解答使用中的困惑。3. 假如这张插图用文字来进行描述的话，带来的结果肯定是一大段复杂、晦涩的文字。这张图片更具有吸引力，占用的空间也不大，而且只要很短的时间就可以明白其中的含义。

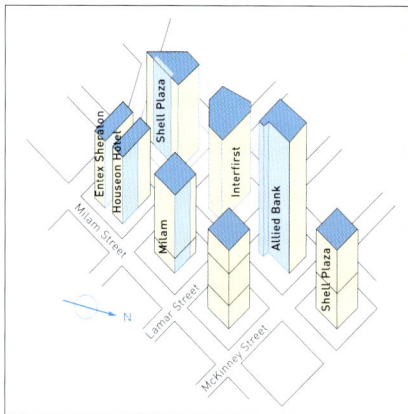

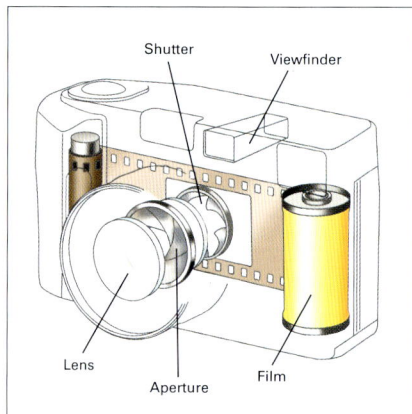

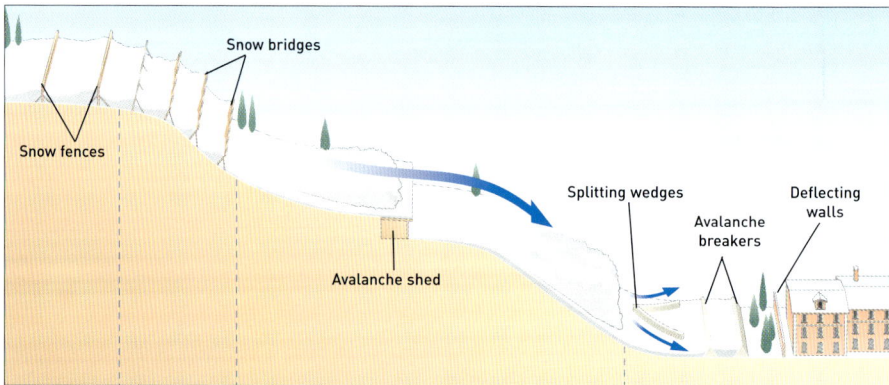

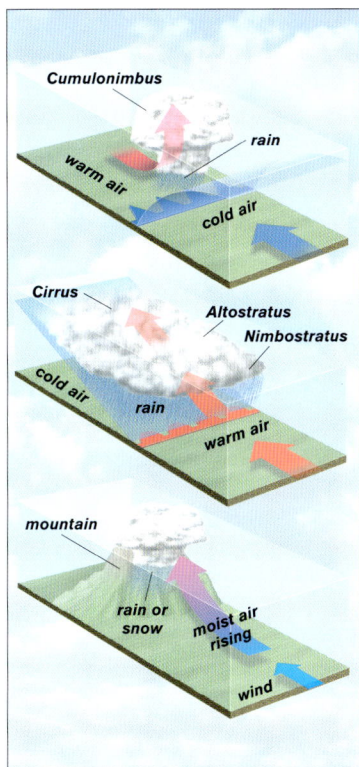

1. 这张清晰、简明的图表传递了大量的信息，如果用文字表达，将是沉闷和复杂的。
2. 图片的作用通过与图表的结合得到增强的良好范例。箭头和图表符号强调了图片中的重点部分。

练习三十一

- 选择一张 6 × 9 英寸的肖像格式页面，设计一张从你家到附近图书馆的地图，标明重要的标志性建筑及公交换乘情况。

- 设计一张反映你所在小镇文化中心的地图，包括教堂和历史建筑、电影院、俱乐部、酒吧等。如果你认为需要的话，可以加图例。

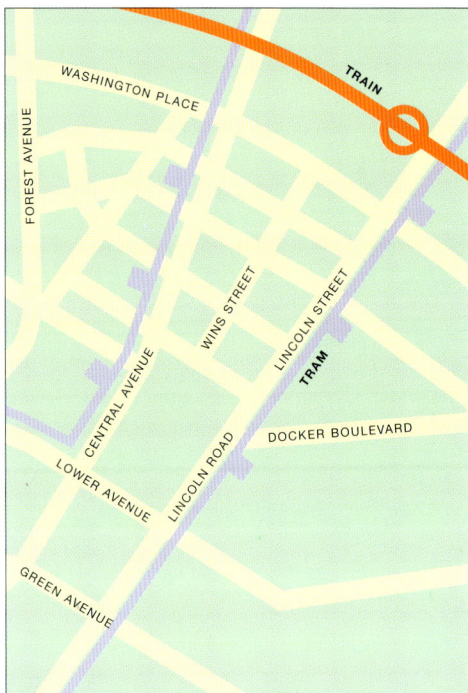

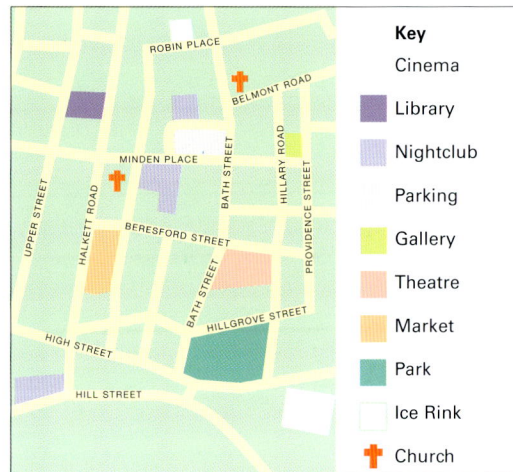

Key

Cinema
Library
Nightclub
Parking
Gallery
Theatre
Market
Ice Rink
Church

练习三十二

- 使用与练习三十一中相同格式的页面，设计一张日常工作（如泡茶、使用书目、煮鸡蛋、缝纽扣或种植盆景）的程序图。

- 设计程序应包含 4 到 8 个步骤，每个步骤占用两个正方框的空间。

- 对每一个步骤进行注释，在 6 × 9 英寸的页面中进行排版，并加上恰当的标题。

Potting Plants

When you buy a new plant, remove approximately one-third of the growth and thin out the branches. This will give the plant a good basis for growth.

The root system will be enormous when new. In spring, reduce the root system by one-third so that the plant does not become pot-bound.

If you wish to transfer the plant into a new pot it is best to do this when root pruning. Allow the foliage to grow quite freely at this stage, this will help the plant to establish itself within the new pot.

Keep trimming the branches to encourage the plant to grow. Sunlight and occasional watering are also required to keep your plant in peak condition.

插图和图表

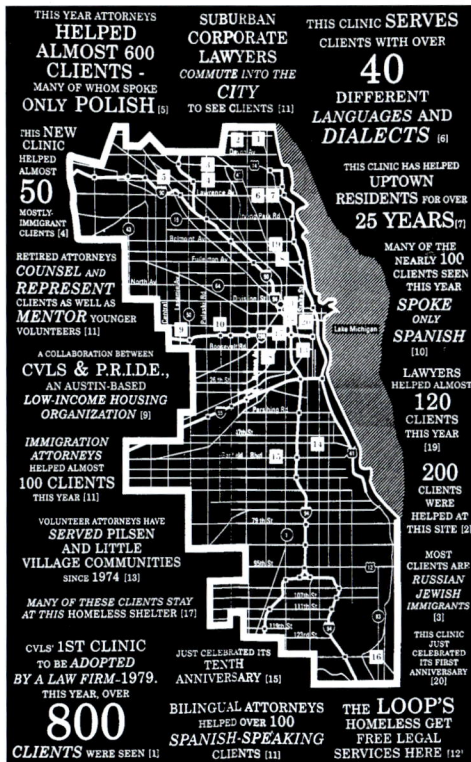

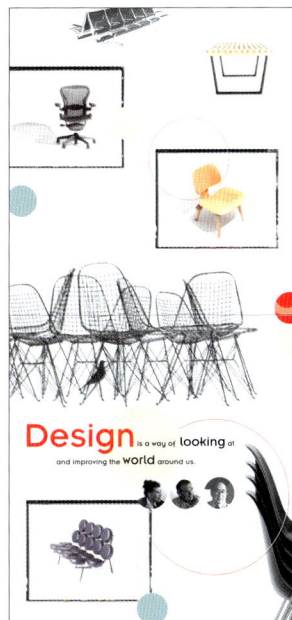

1. 这张复杂的图表有效地传递了信息。字体的不同大小、文字与图表的组合形成了引人注目的设计。

2. 并列的条目可以通过颜色和字体磅值的区别来增强视觉效果。画面右侧的地图简洁而实用。

3. "设计师观察并改善我们周围世界的一种方式",这句话通过这张时尚、平衡、协调的设计得到了完美体现。

4. 将较大的字体放在焦距之外,从而使读者的眼光集中在标题中间的内容上,这确实是一个聪明的主意。

因为我们习惯于在电视和出版物中看到现实生活中真实事件的
照片，因此照片相对于插图而言似乎更真实可信。所以，设计师们
倾向于采用照片使某些信息更可信。

照片
作出决定

使用照片

你可以自己拍摄照片，也可以雇用
摄影师或从成千上万的图片库中搜索你
需要的照片。你还能利用诸如Adobe
Photoshop之类的图像处理软件对照片进
行修饰。

为了更有效地利用照片，你需要懂
得如何构造出具有动感的组合，或者更
重要的是，你需要懂得如何将照片和文
字连成一体。

设计效果： 在处理照片时有各种各样的技巧
可以使用。1. 照片拼贴是传递包含众多要素
的视觉信息的有效手段。2-6. 诸如Adobe
Photoshop之类的电脑软件提供了修饰照片的
多种途径。你可以改变照片的颜色和结构，如
果需要的话，甚至可以进行夸大。7. 简约的
黑白照片通常给人以经典的感觉。8. 不同寻
常及出人意料的色彩增加了照片的表现力。
9-12. 用不同寻常的形状剪切照片也可以使设
计更生动。

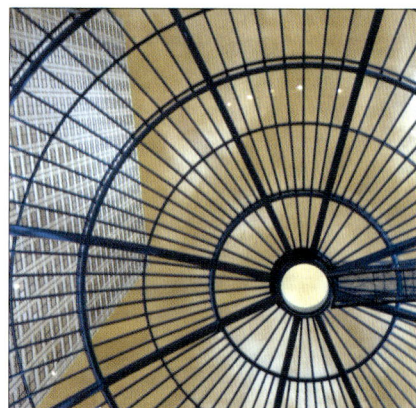

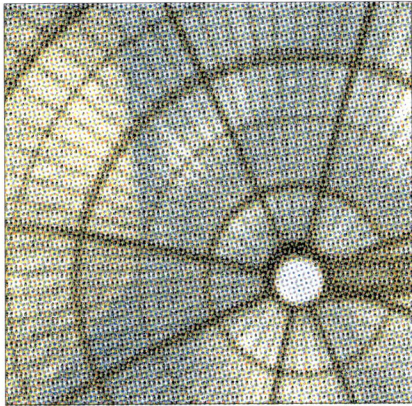

练习三十三

- 选择一张 8.5 × 11 英寸的页面,设计一张双面展开的杂志版面。
- 页面的主题是运动。选择任何一种运动形式,业余或专业的均可,也可以选择包含人和动物在内的运动项目。
- 从尽量多的照片中选择合适的照片。
- 选出六到八张符合主题的照片,这些照片的拍摄角度尽可能多样化,为每一张照片加上说明文字。
- 设计一个网格来容纳图片和文字。
- 利用这些照片和文字,设计出动感、对比鲜明、形式多样的页面。

Young dancers can sweep the floor

The aim of running text around a photograph is to make the words fit with the shape of the photograph without leaving unattractive and clumsy gaps.

Right
You should not insert too much spacing between words to push them out to fit the shape, instead you could add or delete words.

Contrast is an essential element in design. Whether it be in size, color, form, shape, or composition and balance. Here the size difference works well: the cut-out halftone contrasts with the sequence of squared-up halftones.

Practice makes perfect
Running a picture across folds can be hazardous but it can work if you are careful to avoid placing it in such a way that you lose critical parts of the image.

Below left to right
Bleeding off photographs also adds visual interest, but make sure you do not bleed off important areas of the image. Add captions to photographs that are not self explanatory.

练习三十四

- 采用与练习三十三中相同的页面格式和网格,主题变为你的家乡或小镇。
- 选择六到八张具有代表性的精彩照片。可以是建筑物、当地人物风情、街景等。
- 设置照片、形状和尺寸的对比关系。
- 为每张照片写出注释。

Postcard from Brighton

Various formats are used in the photographs on this colorful spread. These, together with the imaginative cropping, make up an attractive composition.

Opposite page
Nam liber tempor cum soluta nobis eleifend option congue nihil imperdiet doming id quod mazim placerat facer possim assum. Lorem ipsum dolor sit amet, consectetuer adipiscing

elit, sed diam nonummy nibh euismod tincidunt ut laoreet dolore magna aliquam erat volutpat. Ut wisi enim ad minim veniam, quis nostrud cipit lobortis nisl ut aliquip ex ea commodo consequat.

Above left to right
Nam liber tempor cum soluta nobis eleifend option congue nihil imperdiet doming id quod mazim placerat facer possim assum. Lorem ipsum dolor sit amet, consectetuer.

Below
Nam liber tempor cum soluta nobis eleifend option congue nihil imperdiet doming id quod mazim placerat facer possim assum. Lorem ipsum dolor sit amet, consectetuer.

A good meal is a work of theater. Ideally, it unfolds gracefully, with Act One setting the stage for things to come. In many households, appetizers may be considered a first course when guests are seated at the table. But in this chapter, we think of them as munchables: tiny morsels that can be enjoyed prior to sitting down for the main act.

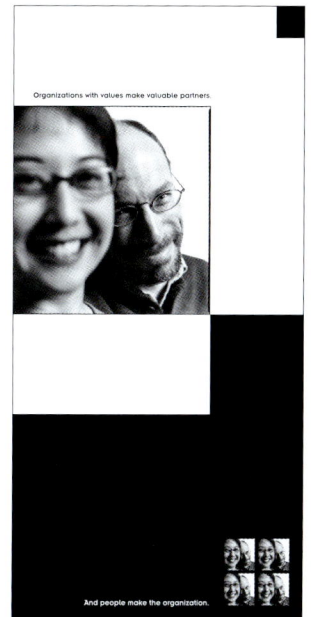

Organizations with values make valuable partners.

And people make the organization.

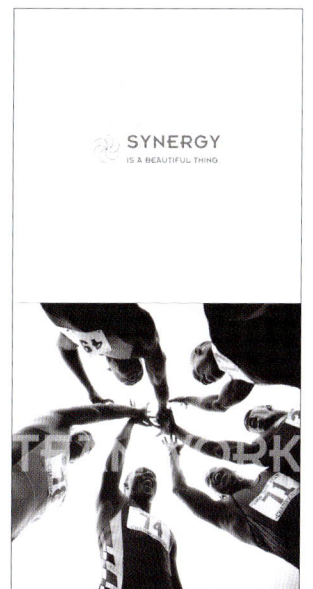

SYNERGY
IS A BEAUTIFUL THING

1. 小照片同文字搭配得很合适。

2. 逼真是使用照片的最主要原因,特别是当表现主题是食物时。

3. 色彩丰富的照片能很好地说明主题,同时也能反映出文字的含义。

4. 使用不同尺寸而内容相同的照片可以增强设计的效果。

5. 富有想像力的摄影师能强化设计效果,像这本小册子一样,增加视觉张力。

　　很少有照片能够按照原样被采用。一方面，如果你使用的照片拍摄格式不同，就不能很好地放置到网格中。另一方面，照片含有的大量细节往往会分散读者对主题的注意力。因此，你需要对照片进行剪切。

照片
剪切和尺寸调整

如何实施

　　构建一个可以容纳你选择照片的大部分内容的网格（见30页）。这会给你关于形状、尺寸和组合方式的最初启发。照片应该充满整个单元格，以达到平衡和协调。

　　剪切会使照片与先前大不相同。剪切的最佳切入点是主题，将与主题无关的部分剔除，并将主题部分尽可能放大。剪切掉你不需要的部分后，将剩下的部分放大至充满网格的单元格。

设计效果： 1. 这张照片包含了过多的细节，分散了观察者对主题的注意力。2. 对照片进行剪切后形成了一张更有表现力的照片。3-8. 这一系列步骤反映了一张简单的照片如何被剪切成不同风格的图片。9. 背景中过多的细节会破坏照片的感觉。10. 放大照片中的某一部分，你可以轻而易举地为照片增添魅力。11. 将背景整个去除，只留下主题部分，可以很好地突出照片的效果，特别是当主题是人物肖像时。

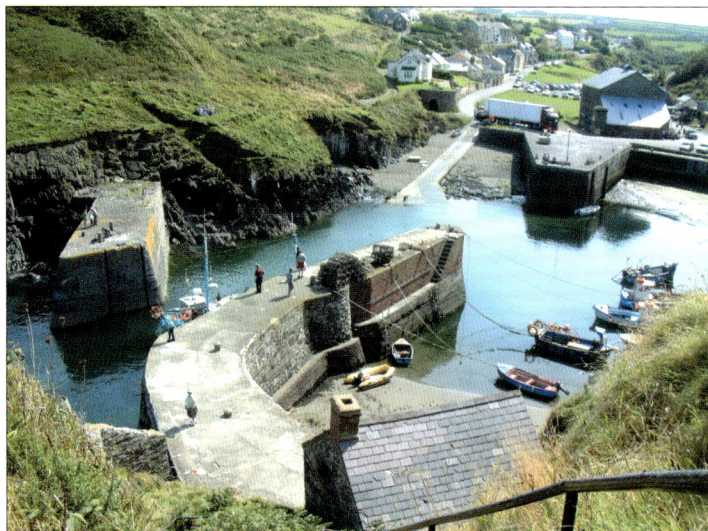

1

2

练习三十五

- 选择一张包含人物在内的运动类照片。
- 采用合适的格式容纳下整张未经切割的照片。
- 逐渐减少照片中包含的人物数量，最后只保留一个人物。

练习三十六

- 选择一张包含几个人物的家庭照片，用合适的形状放置在页面中，并保留所有的背景细节。
- 对照片进行剪切，使人物部分尽可能地扩大。在让人物成为中心的同时也保留部分背景细节。
- 处理同一张照片，但去除所有的背景。

剪切和尺寸调整

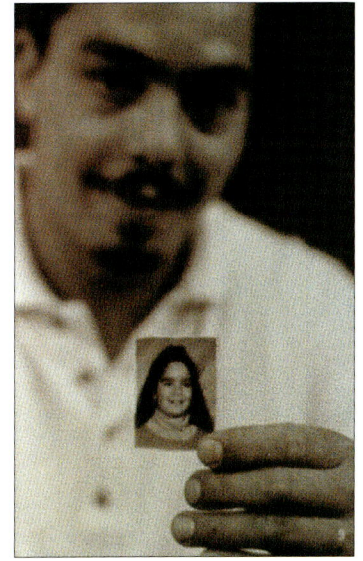

1. 剪切赋予了每一张照片不同的"故事"。
2. 焦距是照片中要考虑的一个重要因素。
3. 照片尺寸大小的对比通常能将目光吸引到照片上。
4. 蒙太奇式的照片能够使设计生动有趣并富有想像力。去除背景后的照片形状更增添了页面的动感效果。

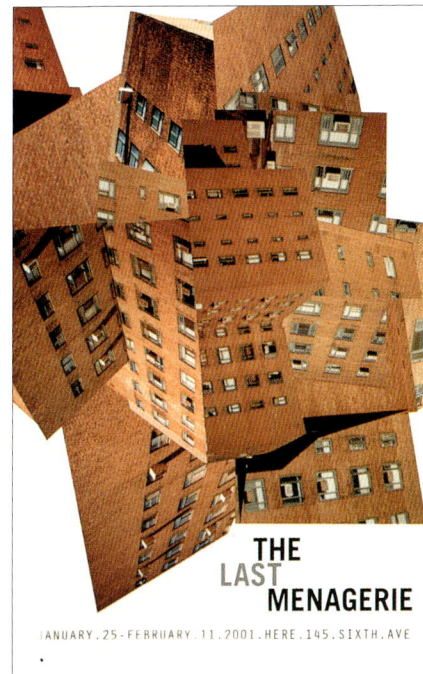

当你把照片和文字组合起来时，有一种很自然的倾向，就是把文字放置在图片上。而这样通常会影响图片的效果。一条最基本的规则是，文字不要放到图片上，或者至少避开图片的主体部位。

文字与图片
组合

状态与位置

尽量选择与图片意境相符的字体。如一张充满激情和力量的照片需要用无衬线的粗体与之相匹配。相反，柔和的图片则需要纤细、精致的字体予以强化，诸如 Garamond 和 Caslon 之类的衬线字体的斜体都很合适。

你把文字放置在何处是非常重要的。仔细分析图片，找寻每一个元素中关于放文字位置的任何可能的启示。

设计效果：1. 这张图片的主体部分包含了多种角度和方位，对此你可以在文字排版中加以回应。2. 图片和文字的角度形成对比可以造就具有动感的页面。3. 图片的说明文字往往可以决定图片与文字的大小关系。4. 是文字重要还是图片重要？ 5. 如果页面空间不够，你或许不得不将文字放到图片上。但必须保证文字不会影响图片的效果，或者掩盖了图片的精华部分。

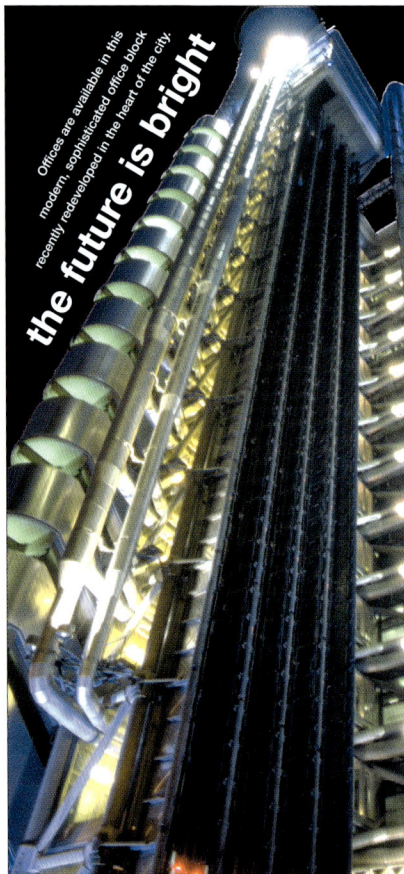

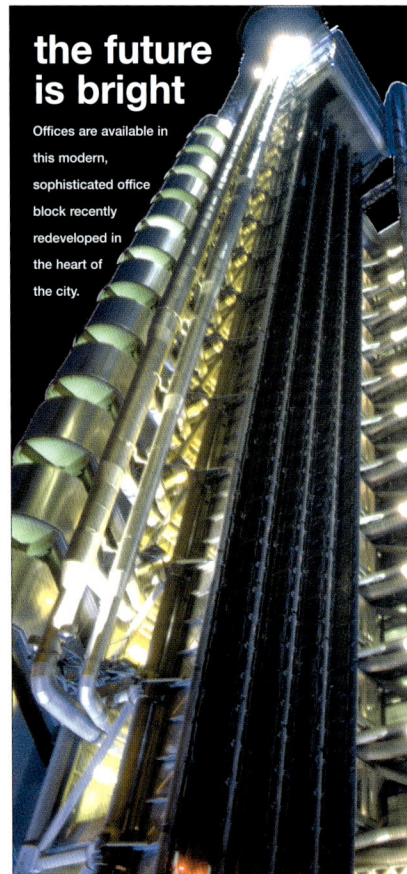

the sky's the limit

Rislin Olympic Stadium

Sport Innovation

Official opening

January 23

in the presence of:

HRH Princess Mary

3

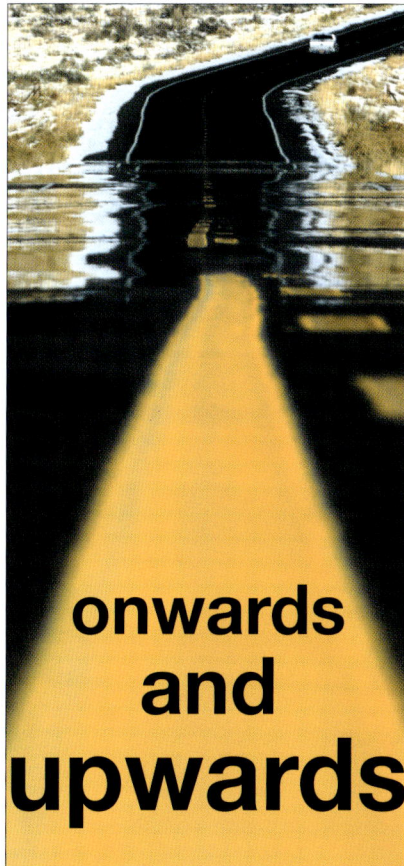

onwards and upwards

4

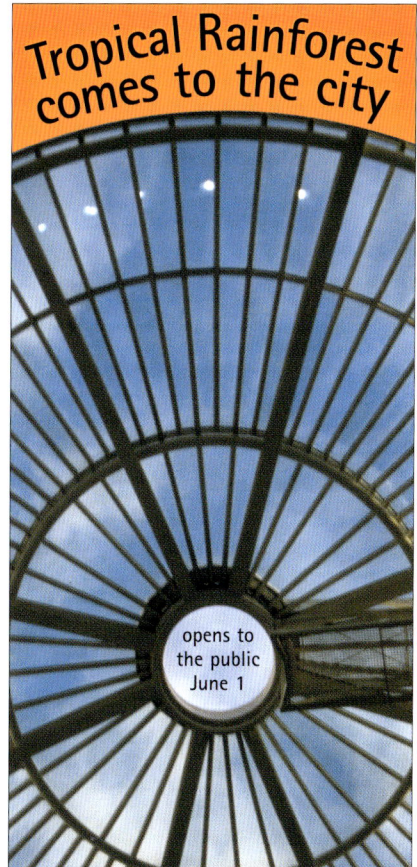

Tropical Rainforest comes to the city

opens to the public June 1

5

练习三十七

- 建立一个8.5 × 11英寸的页面，选择一张运动类照片，并为它加上标题和描述性文字。
- 将文字和图片组合，文字与图片方向保持一致。

Reach for the Skies

Just about any competent skater should be able to learn this trick in a month or two. It takes a while to learn the necessary coordination, but after that it is just a matter of practise. Keep your confidence, practise the move every time you go out skating and you will improve. Don't sweat about falling over: falling is a part of skating, and most likely you're not pushing your abilities if you're not taking a tumble once in a while. Stick with it and pretty soon you'll be flying with the birds – it's a wicked feeling and one worth all your effort. Go for it!

练习三十八

- 使用练习三十七中的文字和图片，对标题和文字进行排版，使它们的方向与图片方向不一致。
- 另选一张照片、一个标题和一段文字。使用无衬线的字体。通过排版使字体、文字、照片和它们的方向相互协调。

- 重复上述练习，并使无衬线字体与照片的意境匹配。

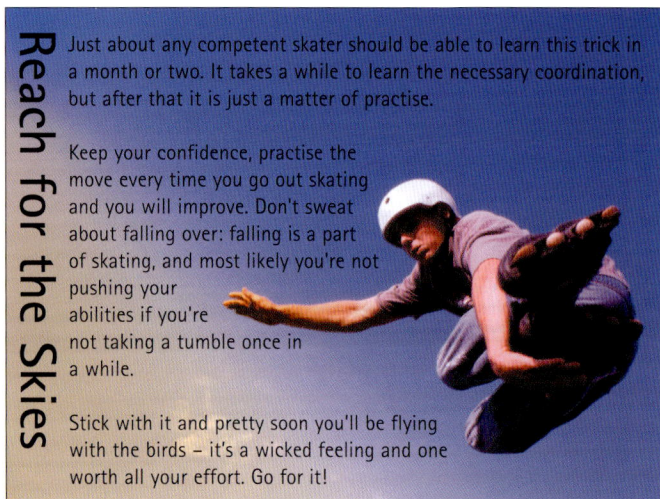

Reach for the Skies

Just about any competent skater should be able to learn this trick in a month or two. It takes a while to learn the necessary coordination, but after that it is just a matter of practise.

Keep your confidence, practise the move every time you go out skating and you will improve. Don't sweat about falling over: falling is a part of skating, and most likely you're not pushing your abilities if you're not taking a tumble once in a while.

Stick with it and pretty soon you'll be flying with the birds – it's a wicked feeling and one worth all your effort. Go for it!

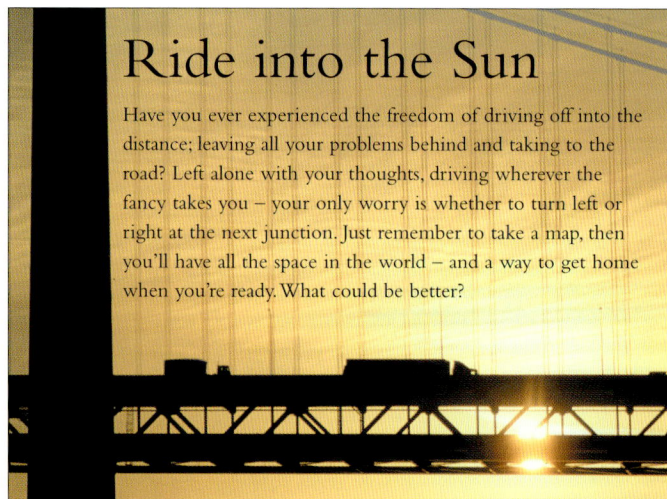

Ride into the Sun

Have you ever experienced the freedom of driving off into the distance; leaving all your problems behind and taking to the road? Left alone with your thoughts, driving wherever the fancy takes you – your only worry is whether to turn left or right at the next junction. Just remember to take a map, then you'll have all the space in the world – and a way to get home when you're ready. What could be better?

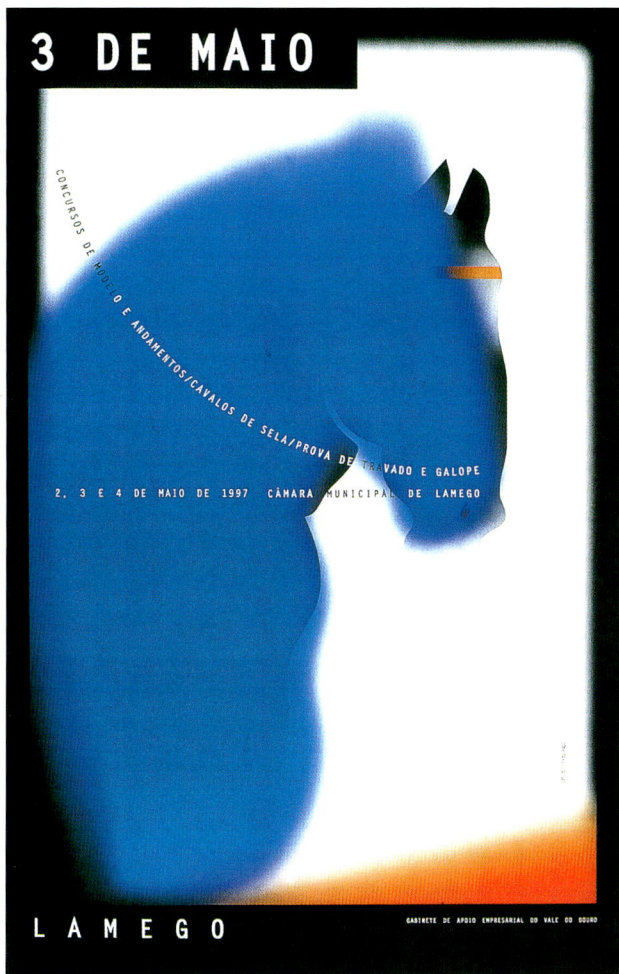

1. 细小文字的特别形状将注意力吸引到了图片的中心部位。其他的元素均在图片边框之外，因此这张优秀的广告立刻就吸引住了人们的眼球，并且异常清晰。

2. 图片和修饰的有趣并置。图片上的字母小心翼翼地放置，以免破坏了图片的整体形象。

3. 充满活力而又色彩丰富的图片和文字很好地反映了设计的主题（男孩夏令营）。

two magpies = 雙喜 double

happiness (shuāng xi)

magpie : The characters for
'magpie,' xǐ què, literally mean
the 'bird of happiness.' A
picture of two magpies facing
each other stands for 'double
happiness,' shuāng xǐ, symbolic of conjugal bliss. The call
of a magpie foretells the arrival
of a guest, good news, or good
fortune. A magpie resting on
a plum branch conveys the
wish 'happiness before one's
brow,' xǐ shàng méi shāo, as
the word for 'plum' and 'brow'

are both pronounced méi.
Magpies also served to preserve the integrity of a marriage,
according to legend. When
a husband and wife were to
be apart for any reason, they
would break a mirror and
each take half. If the wife was
unfaithful, her half of the mirror
turned into a magpie that
flew back and informed her
husband. Consequently, an
image of a magpie is often
placed on the back of a mirror.

double happiness · 171

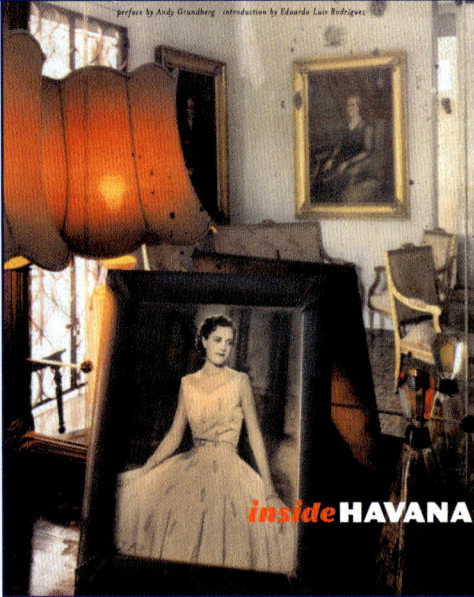

preface by Andy Grundberg   introduction by Eduardo Luis Rodríguez

*inside*HAVANA

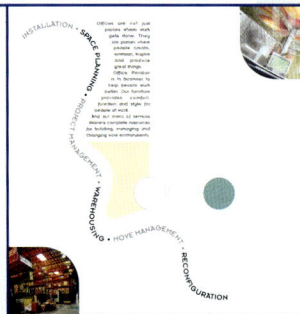

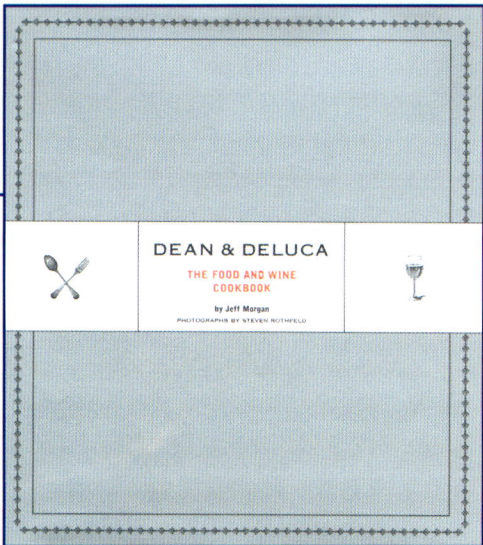

第二部分

设计项目及类别

现在,你已经获知了基本的设计理论,到了将它们付诸实施的时候了。

无论你的设计对象是什么,每一个设计项目都要经历如下三个阶段: 首先是简要介绍阶段,在这个阶段中,你与客户一同商讨设计项目要达到的目的和效果。接着是设计酝酿阶段,简明扼要地写出想法可以帮助你理清思路。最后是产生成果阶段,在这个阶段中,设计师与设计小组之间的沟通同设计师与客户之间的沟通一样重要。

本部分的内容主要涉及的是第二阶段——创意阶段。设计项目被划分成各种类别,每一种类别的要求各不相同。接下来的章节将会探讨其中的一些类别,并用范例分析和专家示范的方式说明创意过程。同样也提供了帮助你提高设计水平和进行自我评估的练习题,从中你可以逐步建造你的知识大厦并获取自信,最终成为一名富有创意和激情的设计师。

DEAN & DELUCA

THE FOOD AND WINE
COOKBOOK

by Jeff Morgan

PHOTOGRAPHS BY STEVEN ROTHFELD

概　　述

为什么需要这项设计？

这个问题要澄清设计的目的和目标市场（例如，是需要一个推销某种产品的促销工具，还是仅仅为了传递某些信息）。就此达成共识后，你与客户可以讨论决定需要的特殊条件和其他步骤，如市场调查等。

你打算做什么？

这个问题的答案明确了你将如何达到设计目的。何种设计风格和视觉效果与目标吻合？设计将打印出来还是在屏幕上显示，或者两者都用？需要进行何种视觉上的研究？需要摄影师、插图师或广告文案撰写员协助吗？

设计何时需要？

设计完成的最后期限将直接影响设计和最终成果产生的日程安排。与这个项目相关的三方（客户、设计师和设计小组）均应清楚地知道设计的日程，并已经就各个项目的实施达成共识。客户负责的是设计的预算、文字和目的；设计师负责视觉方面的效果和设计的专业技能；设计小组则保证设计项目按时按要求完成。但在客户和设计小组之间进行沟通的是设计师本人。

预算有多少？

很少有预算是无限制的，而可供设计项目使用的费用对作为设计师的你及你的设计小组均有影响。充足的预算让你可以自由地使用附加色彩和效果，如层叠、透视、浮雕或者图案装饰等。但是，有限的预算也不一定意味着设计效果会差。想像力丰富的设计师和富有进取的设计团队可以通过对色彩和页面等元素的合理使用，使有限的经费发挥最大的作用。优秀的设计师总是能使手里的资源达到最佳的效果。

在简单介绍阶段，你和客户都要确信了解了对方的想法和意见。客户必须清楚地讲明他或她需要什么，而你必须让客户明白你打算怎么做。不要忘记记录下你们已经达成的共识。

- 假设你受邀与一个客户讨论设计一本关于教育的小册子，这本小册子是与两个系列的视听材料一起出版的。这两个系列的视听材料分别是"睡眠的秘密"和"在你的梦中"，都涉及到探索睡眠中无意识的世界。

- 列出一张你打算询问的问题的清单，以确保你的设计达到预先设想好的目标。
- 请同事或朋友假扮客户，和你一起进行练习。讨论设计项目（保证你的问题都得到了回答）并了解客户是否有其他你未涉及到的特殊要求。

- 不要忘记写下你们通过讨论达成的共识。同时要求客户也这么做，并对照你们的记录是否存在差别。
- 下面的范例将教你如何入手。

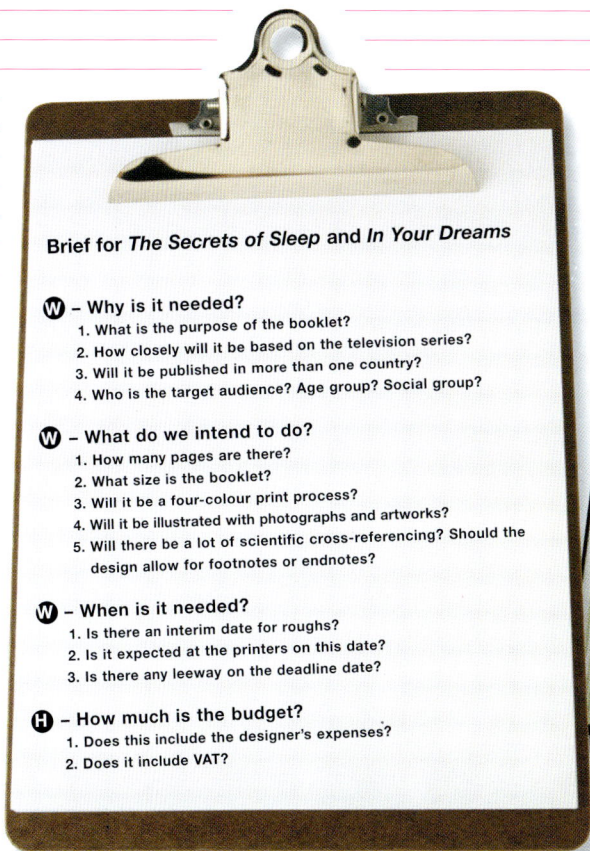

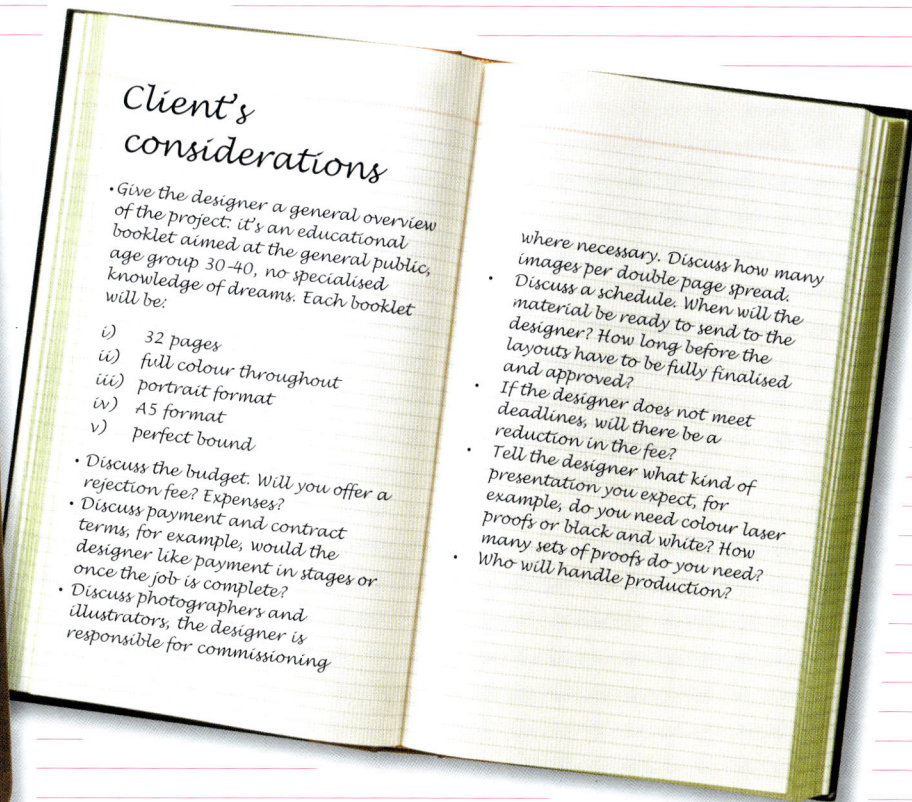

Brief for *The Secrets of Sleep* and *In Your Dreams*

W – Why is it needed?
1. What is the purpose of the booklet?
2. How closely will it be based on the television series?
3. Will it be published in more than one country?
4. Who is the target audience? Age group? Social group?

W – What do we intend to do?
1. How many pages are there?
2. What size is the booklet?
3. Will it be a four-colour print process?
4. Will it be illustrated with photographs and artworks?
5. Will there be a lot of scientific cross-referencing? Should the design allow for footnotes or endnotes?

W – When is it needed?
1. Is there an interim date for roughs?
2. Is it expected at the printers on this date?
3. Is there any leeway on the deadline date?

H – How much is the budget?
1. Does this include the designer's expenses?
2. Does it include VAT?

Client's considerations

- Give the designer a general overview of the project: it's an educational booklet aimed at the general public, age group 30-40, no specialised knowledge of dreams. Each booklet will be:

  i)   32 pages
  ii)   full colour throughout
  iii) portrait format
  iv)   A5 format
  v)   perfect bound

- Discuss the budget. Will you offer a rejection fee? Expenses?
- Discuss payment and contract terms, for example, would the designer like payment in stages or once the job is complete?
- Discuss photographers and illustrators, the designer is responsible for commissioning where necessary. Discuss how many images per double page spread.
- Discuss a schedule. When will the material be ready to send to the designer? How long before the layouts have to be fully finalised and approved?
- If the designer does not meet deadlines, will there be a reduction in the fee?
- Tell the designer what kind of presentation you expect, for example, do you need colour laser proofs or black and white? How many sets of proofs do you need? Who will handle production?

标识可以是明确反映主题的抽象或具象的符号和字母。某些标识仅仅包含抽象或具象的符号，如壳牌的标识。然而，严格说来，标识指的是整个标识中的字体。这也是我们在本部分中要讲述的内容。

标识与信笺

标识

成功的标识字体有三个标准：1. 设计要反映出机构的类别；2. 标识应该简洁明了；3. 可以使用单色或多色印刷，并且可以适合从非常小（如名片）到非常大（如广告招贴）的不同尺寸需要。

从某种意义上讲，你最好从设计黑白标识开始。不要忘记你可以利用色调的变化，因为它可以为你提供选择范围很广的灰色。当你对设计的黑白稿满意之后，你可以轻而易举地为设计加上颜色。

标识字体的选择要依据你希望通过它们传达的信息来确定。作出选择后，你需要审视这些字母放在一起是否协调。最佳的办法是选择一种你最喜欢的字体，然后通过调整字间距使组成标识的字母达到视觉上的平衡。你或许要改动某些字符才能达到这一目的，但不要改动太多，因为改动过多可能导致文字难以辨认，或造成格式的不协调。

刚开始着手时，先将文字放到一个简单的图形中，如方形、圆形或三角形均可。将字母调整为1英寸高的黑色字体，背景用白色。然后你可以试试黑白翻转，或者可以试着对背景形状进行改变。前面叙述过的关于字体选择、黑白翻转和色彩使用的相关技巧都可以派上用场。

信笺

设计好标识后，你就可以把它运用到各种文具中，如信笺顶部、印有敬语的纸条和名片等。还是从信笺顶部入手吧，因为这将决定其他用途中标识的使用风格。收集一些其他公司的信笺，并将它们放入文件夹以供参考，这是个不错的主意。

信笺顶部几乎都有固定的格式。信笺通常会在折叠两次后塞入信封，因此可以分三部分考虑：上面三分之一包含公司的标识、名称和地址，这些要素应互相协调。同时，你还要注意是否有重要的信息放到了折叠部位。比较好的解决方案是将地址放在左边（实际上，如果使用的是窗口式信封的话，这样做显得更重要）。标识的位置则要与地址协调，尽管最后的放置地方要视你希望形成的总体效果而定。

你也应该和客户商讨哪种字体适合放在信笺顶部。如果有其他的附加信息要加入（如不同的分公司或分支机构的列表），可以将它们放到信笺底部或侧面。最后，记住要留出空白部位以供书写。

其他的辅助手段，如烫金（金属材质印刷到页面上）、压凹凸印（通过压力或粗糙的颜料增加页面的凹凸感）、浮雕（凹进或凸出纸面）等都可以使用在信笺设计上。

- 从下列公司中选择一家作为你的对象,列出一系列标识字体设计的步骤:

  嗅觉——一家以鲜花为销售对象的连锁商,与植物类产品竞争市场。

  最佳效果 (Optimum) ——一家照相机和光学器材生产商,以生产小型单镜头照相机而与佳能和尼康进行竞争。

  贾科尼亚特提 (Giacometti) ——一家提供传统意大利食品的意大利饭馆,它希望设计有现代的感觉。

- 标识要反映出公司的产品、服务及特性。要求自己只着手选择标志字体和基本的形状,不要添加其他任何元素。为你的设计写出书面的注释,包括解释创意产生过程及评价设计的潜力。

- 将标识用到信笺、名片和印有敬语的纸条中,并加上公司地址。尝试各种组合,然后将它们打印出来,比较哪一种效果最好。

- 下面的创意将教你如何着手。

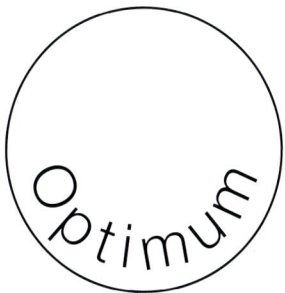
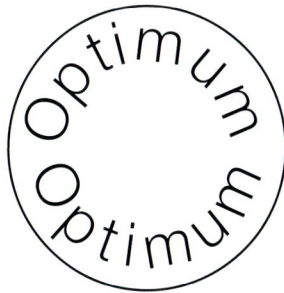

标识
范例解析

Teatro Bruto 的设计，反映了客户希望进行的戏剧性尝试。Bruto 意为粗糙的，未经打磨的，而实际上公司的项目是尝试性的，但却非常专业化。橡皮图章似的字体与程度不同的压缩结合，传递出一种紧张、不舒服的感觉。设计的最终效果得到了体现。

Estudio de Opera do Porto 是音乐学校的大多数签约歌手的经纪公司。它的标识选择了字母 "0" 代表歌手张开的嘴巴，而引号则代表了歌剧中的其他元素，如歌词和剧本。

1. 从尝试各种字体和颜色出发，寻找哪一种字体最适合转变为橡皮图章式字体。
2. 凑近些看，单个的字母中可以看出类似敲印章的过程中墨水不够的情况。因此，简洁清晰的无衬线字体在这里比较合适，字母一部分的缺失不会影响正常的阅读。
3. 设计出的标识非常醒目，并且准确反映出了客户从事的行业。

teatro bruto

TEATRO BRUTO

TEATRO BRUTO

TEATRO BRUTO

TEATRO BRUTO

TEATRO BRUTO

TEATRO BRUTO

TEATRO BRUTO

1

2

TEATRO BRUTO

3

专业范例

4. 引号非常粗黑，并且足够将字母"O"囊括其中，以此划分出标识与页面其他部分的界限，避免与页面的其他部分混合。
5. 清晰的无衬线字体，表现出了正在歌唱时的嘴的形状。
6. 这些文字清楚地传达了标识包含的信息。
7. 设计出的作品虽然复杂，但却很精致，三个部分和谐地组合在一起，简洁明了地勾勒出了反映的对象。

4

O

ESTÚDIO
DE ÓPERA
DO PORTO
CASA DA MÚSICA

5

6

8

TEAOLOGY

9

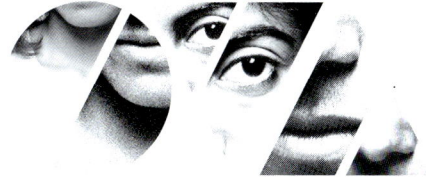

725 N. Milwaukee St. Milw. WI 53202 414 276-6363

10

11

"O"
ESTÚDIO
DE ÓPERA
DO PORTO
CASA DA MÚSICA

7

8. 设计突出了三个要素：全球性、数字化、"Allavida"的打头字母A的粗体，均反映出了该机构的风格和重心。
9. 这个看似普通的黑白标识，是为一家健康茶叶生产公司设计的。在字母L上加上了茶叶形状的图形，指明了标识所包含的意义。
10. 这个具有"手写体"风格的标识是为一家美术馆设计的。随手书写的美术馆名字、打印的地址和不经意的排列，表现出了艺术的自由和创造魅力。
11. 标识字体拼写为"VOA"，采用了部分图片拼接的形式，暗示了不同的内容和含义。

时事通讯是杂志和报纸的混合物，通常是由大中型公司、俱乐部和休闲机构印刷发行的。它为新闻、社会事件、机构内员工动态及主要事件提供了告知的平台。

时事通讯

一般来说，时事通讯不像报纸和杂志那样伴随有时间的紧迫性，充足的时间让你可以实施良好的创意。设计的施行通常有一套基本的模式：首先是预算，通常主要内容只用一种颜色，而封面可能有两种。如果有诸如此类的限制，就采用字体、线条、渲染等手段增加背景的变化。如果封面和封底能用两种颜色的话，你可以用其中的一种颜色的不同色调来反映图片，而另一种颜色则用于标题。

第二个要考虑的是版式。这也许会受后续工作（如需要塞进标准信封，或需要控制分量以免邮寄费超标）和需要排版的文字多少的影响。在页面中塞进过多文字，往往在视觉上令读者不是很舒服；因此要尽量保留大量的页面空间，并通过改变图片尺寸和字体增加页面的变化。如果有许多文字要编排进去而页数又不能增加，则可以选择无衬线的小字体或者较窄的字体帮你把文字排进页面。

在设计时事通讯时首先要做的是分拆页面。将它们分成若干部分：图片、新闻、运动、员工动态等等。在考虑建立何种网格时，它将给你很好的帮助。图片的列宽一般较宽，而新闻、运动或员工动态等则更适合放到窄一些的列中。

建立网格后，在页面的各个部分中进行尝试。线条和标题字体可以帮助你区分各个部分，并且使页面清晰明了。在正文部分，遵循本书第一部分中提到的原则，你可以通过改变字体和格式强调某个专题（如从罗马字体变为斜体），或者可以加入

条块和色彩。它的整体风格应与组织机构本身的特征相符——例如新潮现代或传统保守。

在建立了网格并确定了整体风格之后，需要为以后的用途设计好固定的样式。公司也许会自己出后继的刊物，这时你的角色是监督者和咨询者。如果固定的样式兼容并蓄，发生问题的几率将会很小。

appearance
summer issue

appearance
summer issue

- 你受邀为一家房地产开发公司设计一份时事通讯。这家房地产开发公司的业务涉及建筑设计、建造、内装潢设计和绿化。时事通讯的名称为"表现（Appear-ance）"，目标群体是介于25到40岁之间的人群。这份刊物看上去应非常现代。

- 设计一份刊头为"夏季号"的时事通讯封面。
- 寻找合适的照片，设计一张打开的页面，其中包括一篇包含适当小标题的配图文章。
- 设计另一张展开的页面，反映员工离开和加入公司的状况。可以采用个人的小照片，并为以后的使用准备一些备份。

- 请一位同事分析一下你努力的成果，并讨论设计的优势和弱势。
- 下面的设计对你会有所启发。

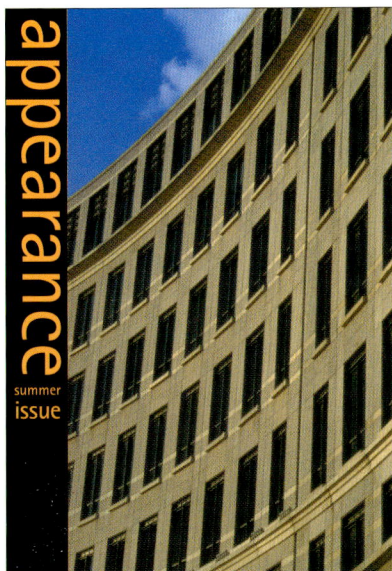

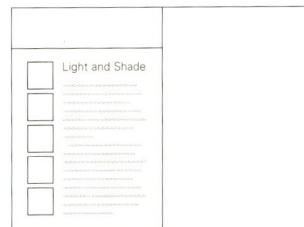

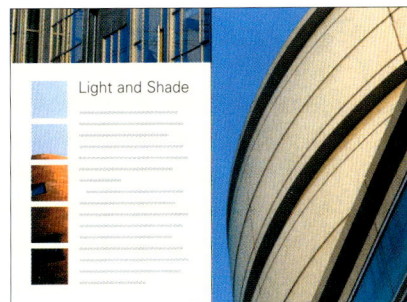

有时你完全可以在设计一份时事通讯时勇敢地进行尝试，但这大部分得依据客户的需要而定。设计的风格应该反映出组织机构对自身的定位，明白他们需要传达出多少信息，了解他们的预算是多少。

时事通讯
范例解析

这份时事通讯是专为 FIT（冰岛平面设计师协会）的成员设计的。它的目的是以有趣而时尚的风格发布工业新闻和声明。至今为止，它打破了平面设计和排版的传统观念。鲜明的色彩、丰富的形状、空间、字体及丰富多彩的排版风格等，均为设计带来了生机勃勃的活力。

在时事通讯中仅仅使用印刷和排字元素进行设计，是一次非同寻常的冒险。然而在这里，这种冒险取得了成功，因为它实现了客户所在的组织机构想要达到的目的。

1. 注意标题的字体是如何排列的。
2. 线条非同寻常的大胆运用，限定了刊头的范围，并平衡了页面，起到了很好的作用。
3. 文字变成了页面的底纹，而不是传递信息的载体。
4. 为一本科技类时事通讯所作的简洁有效的设计。
5. 色彩条块、变化的排版风格及行列较短的标题字体，实现的组合是一张生动活泼的页面。
6. 这张页面表现了对空间的充分利用——在刊头和页边没有地方被浪费。垂直排列的刊头、被切割的刊头及左右两边的对齐，均发挥了框定正文的作用。

专业范例

小册子包括各种各样的推销材料和信息发布资料，涵盖了从艺术展览到健康、安全等的广阔范围。因为有着诸多变化，公众开始期待看到激动人心的设计，而设计师不得不满足公众的需求，虽然不一定要有损于信息的清楚传达。

小册子

与时事通讯一样，第一步是在分析内容的基础上建立网格。如果小册子的首要目标是作为推销材料，那么重点就应放在图片信息（照片和插图）上，并辅以说明性文字。这些文字无非是反映价格的数字，或者是对物品的详细说明。如果文字中包含了所有这些要素，那么一个包含小单元格的弹性网格将非常合适，因为它可以让你把数字和各种大小的图片及小块的说明文字结合起来。而较宽的单元格则适合详细的说明性文字。

在其他类型小册子的设计中，图片和文字的比率差不多相同，因此建立的网格不那么复杂。图片需要注释，而注释采用狭长的排版方式看上去比较舒服，为此必须在网格中设计较小的单元格。

在确定了网格并且收到了来自客户的所有资料后，在几张展开的页面上进行尝试，看看文字（包括注释）与图片或插图的搭配情况。你可以很快地在此基础上进行调整。有时向客户展示几个同一张页面的不同设计版本，是一个很好的方法，因为它可以成为进一步商讨的良好开端。

尽管小册子的初衷是传递信息，并且文字多于图片，但你可以尝试改变页面的格式。在处理文字时这或许较难办到，因为你要避免字体和尺寸的过多变化，但你同样也要注意图片与文字的对比。你可以通过尺寸（如大图片与小图片的对比）或色彩的变化（如在主要的彩色图片中插入黑白图片）来达到目

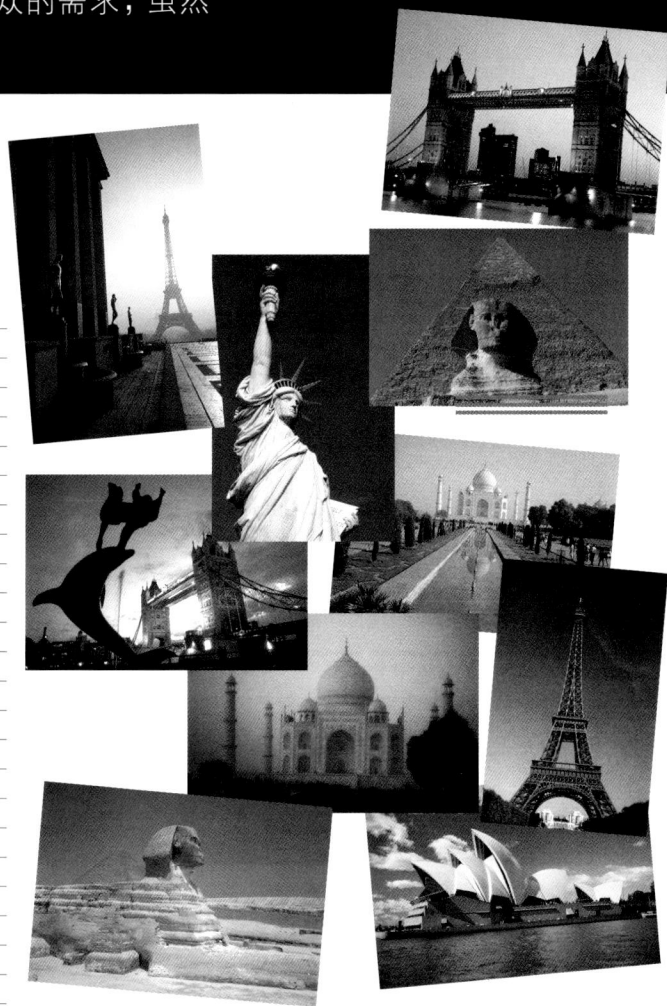

的。对图片进行这样的处理需要剪切图片的基本技巧（见78—79页）。用各种元素将页面组织起来，同时避免过于花俏的炫耀，这些都是设计好小册子的关键。

- 为一次图片展的宣传手册设计3张展开的黑白页面，在3张页面中至少包含8张图片。
- 为每张图片撰写恰当的注释。
- 然后设计出封面并加上恰当的标题和图片。
- 与一个同事一起分析设计效果，并讨论设计的优势和弱点。
- 下面的范例将帮助你展开设计。

Places of Culture

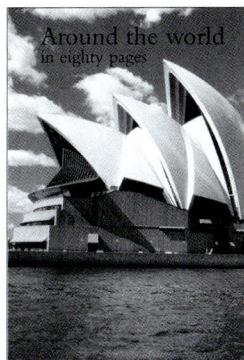

Around the world
in eighty pages

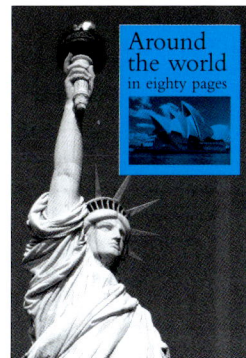

Around
the world
in eighty pages

Places of Culture

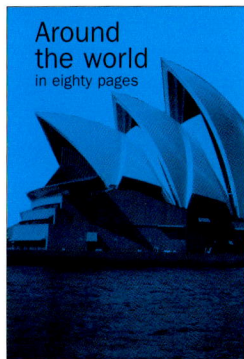

Around
the world
in eighty pages

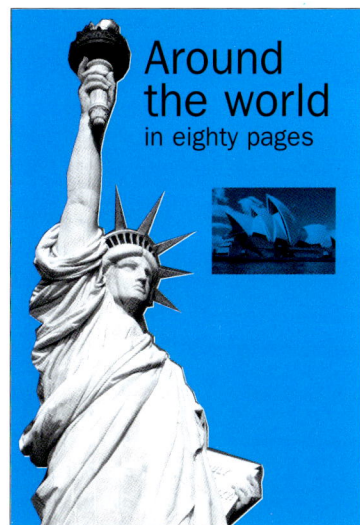

Around
the world
in eighty pages

作为推销材料的小册子需要色彩丰富、图文并茂。在这一页中，为宣传一家提供流程解决方案的公司而设计的小册子，是一个文字与图片结合的优秀范例。小册子的标题"效率"通过图片得到了强化，而图片又选择了最能体现"效率"的物品——钟。

小册子
范例解析

这张页面是引人注目的。在页面中有着强烈的对比——右页中是文字而左页却整个被图片占据。空白部分的使用也是非常慷慨的，它使得视线能够在不同信息之间轻松随意地游走。

各种宽度、尺寸和风格的变化使文字的排版富有想像力。左对齐、右对齐和居中的排列方式都得到了很好地运用。最终的效果是吸引人的，而且不会太繁琐。

1. 小标题和正文常用的左对齐方式，清楚地传达出了信息。同时也平衡了精致图片形成的效果，避免页面过分繁杂。
2. 将文字用作图片的有趣尝试。在"in"的重复字句明显少于在"out"上的字句，以此强化了高效率的生产和销售信息。
3. 跑表的部分图像被拖到了右页中。这有两个目的：首先，这部分图像可以作为一个标准，在上面书写"效率"一词（以跑表自身的一部分来作为标准是多么恰当啊）；其次，拖动图像后形成的空白让这张满当当的页面略显放松；而相应地，这部分图像又填补了右页中原本会出现的空白。

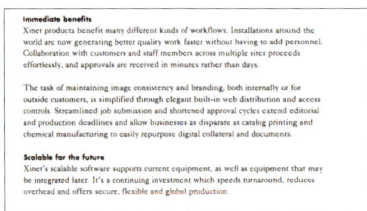

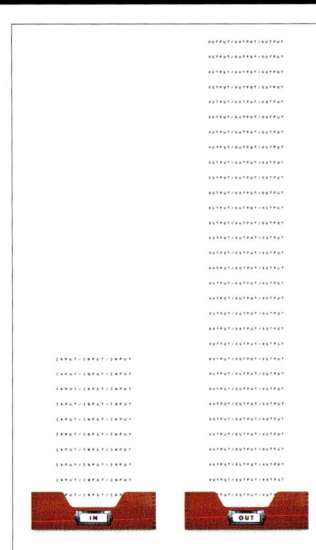

专业范例

4

5

6

7

4. 这张页面设计的重点在于宽阔的上下页边距和垂直的粗线条，这些线条构成了页面的左右边距，并划分出了图片和文字的范围。虽然图片越过中缝延伸进右页或许是个不足，但标题文字却是位于左页图片正中的。

5. 拥有生动丰富的图形和图片的结合，使这本小册子的封面实现了很好的平衡。

6. 这张图片之所以立即引起注意，是因为图片放置的角度非常特别。图片在页面中摆放的位置，产生了三维的效果，并使之浑然一体。

7. 展开的两张页面稍微有些怪诞。这种感觉的产生，依靠的是有限的几种色彩。刊头通过红色条块的衬托跃然纸上。

杂志

和时事通讯及小册子一样，你需要在着手设计前分析目标群体和杂志的目的，因为不同类型的杂志需要不同的方式。内容决定网格形式：一些杂志的网格排列紧凑（大多为文字量大且集中的杂志），而另一些（重点在图片要素上的杂志）则较为松散。

与前面提到的其他种类的设计相比，在杂志的设计中，客户提供的材料对设计的成功尤为重要。因为客户提供的材料（包括图片和文字材料）直接关系到页面的外观，而页面的外观和感觉完全根据客户的需要设计。例如，严谨的学术杂志通常是文字占主导的，因此你应该专注于字体的大小尺寸和行间距，避免造成过多的单词拆分而影响阅读。可以使用衬线字体形成杂志严谨、传统的风格。而无衬线字体作为标题，与正文形成对比，又能使页面具有现代感。

读者对杂志一般倾向于跳读，而不是一页接一页地翻阅整本杂志，这即是加入丰富视觉元素的重要原因。尽量形成页面间的变化——这意味着如果一张页面是宁静且文字较多的，那么下一张页面就应该具有丰富的视觉元素。刚开始时，可以浏览市场上丰富的杂志资源，以获得满足特殊需要的灵感。优秀的设计范例将展示出图片和文字是如何生动地组合起来的。

ART**WORKS**

art works

art works

artWORKS

works

art

art works

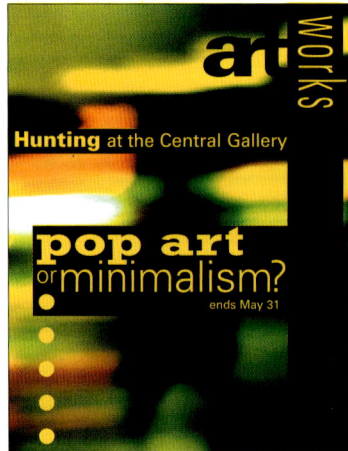

- 你受邀为一份艺术杂志的重新设计确定基调，杂志的名称可以从以下几个当中挑选：现代艺术、当代艺术、艺术表现、艺术品。主要人物是艺术家亚力克·亨特。

- 用你选择的标题设计一个刊头。
- 然后设计一张封面，包含刊头和如下信息：
  亨特在中央美术馆
  通俗艺术还是极简主义？

结束于 5 月 31 日

- 构建一个包含五到六列单元格的网格，然后选择最多三种字体，分别用于正文、引文、正文中部分需要突出的文字、标题和注释。

- 设计至少两张关于亚力克·亨特（或者你自己选择的艺术主题）的展开的页面。注意文章的节奏，在两页中变换文字和图片的比例。

- 下面的创意会帮助你着手设计。

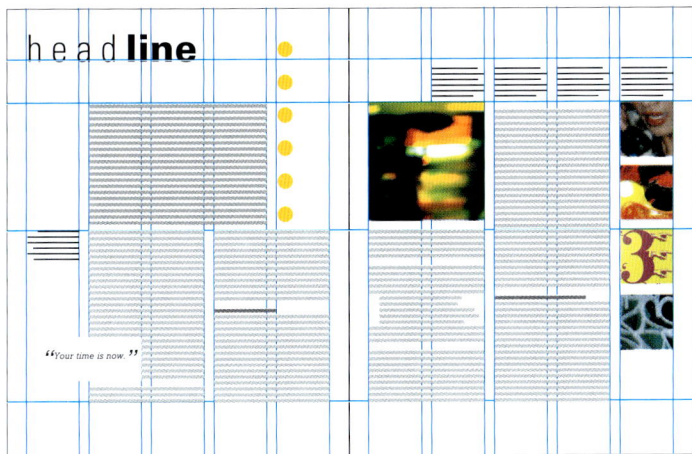

h e a d **line**

"Your time is now."

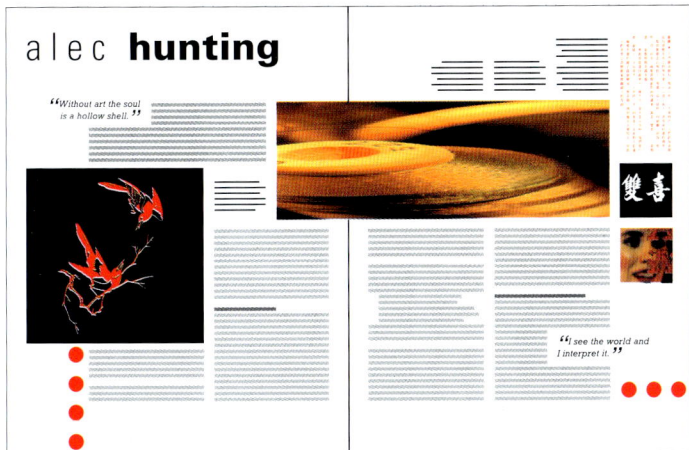

a l e c **hunting**

"Without art the soul is a hollow shell."

"I see the world and I interpret it."

new w a v e

"Just master it. Believe you can and you will"

《视觉和听觉》是一本起源于 20 世纪 30 年代的严谨的电影杂志。电影产业处于时尚和革新的前沿，因此任何以此为主题的杂志均要反映出这一特点。

杂志
范例解析

英国电影学会出版发行了《视觉和听觉》。更新杂志的风格，使之与设计创新、科技创新相协调在这份历史悠久的成功杂志中占据着重要的地位。这些新近的设计尝试引入了新的有趣的元素。

1. 具有现代感的无衬线字体称为"Bruener"体。小小的"&"符号被加粗，以便与刊头中其他的文字相匹配。
2. 标题通常采用无衬线字体，为杂志增添现代、符合潮流的感觉。同时文字的可读性也没有受到影响，因为文字均采用了衬线字体。
3. 杂志设计依靠视觉的变化达到丰富多彩，而颜色条块的运用是达到这一目的的有效途径。它们同时也可以搭建出页面的结构，区分不同的内容。杂志通过缩小字体并在同一单元格中设置两栏文字，成功地将回顾影片的所有细节都包含了进来，这也是对空间极为有效的利用。
4. 这张展开的页面展示了丰富的视觉信息。除了图片大小尺寸的对比外，颜色条块和网格的灵活运用都是杂志设计中的好方法。
5. 标题字体显示了对垂直版式的恰当运用。粗重的线条为文字和页面提供了平衡。
6. 在正文中插入的引文保持了正文的条块结构，并且突出了引文。在注释设置中，形成字体风格的对比（这里是无衬线粗体和衬线字体间的对比）也是一种值得尝试的方法。

Sight&Sound

The Monthly Film Magazine/November 2002/£3.25   1

Interview

Premium Bond

Edward Lawrenson: Were you surprised when you got the call to do the movie?
Lee Tamahori: I was, because I thought I'd have been the last person they'd ask. I was doing another hard-edged, visceral movie in LA which had just fallen apart. My agent phoned and I didn't hesitate. First, I wanted to do a movie my kids could watch because   2

Two Men Went to War

United Kingdom 2002

Director	**Music Supervisor**
John Henderson	Robin Morrison
Producers	**Music Producer**
Ira Trattner	Michael Paert
Pat Harding	**Production Co-ordinator**
Screenplay	Claire Griffin
Richard Everett	**Executive Producer**
Christopher Villiers	Tony Prior
Director of Photography	Fireworks Music Ltd
John Ignatius	**Music Editor**
Editor	Michael Paert
David Yardley	**Music Recordist/Mixer**
Production Designers	Gerry O'Riordan
Sophie Becher	**Soundtrack**
Steve Carter	"Run, Rabbit,Run", "(We're
Music/Music Conductor	Gonna Hang Out) The
Richard Harvey	Washing on the Siegfried
	Line" – Flanagan and
©Two Men Went to War	Allen; "Turn Your Money
Partnership	in Your Pocket" – Jimmy

3

little dangerous handing out guns in a bank" asks Moore.

After this things dip slightly, with Moore going for a soft target: the Michigan Militia. At one of these meetings, a shooter – a real estate negotiator when not in fatigues – asks Moore: 'Who's going to defend your kids? The cops? The Federal Government?' Moore raises his eyebrows and plays to the gallery with a lame joke "Do you take one of those guns with you when you're negotiating real estate?"

It's a rare low point. The Michigan Militia was created in the wake of the Ruby Ridge massacre in Idaho in 1992. At Ruby Ridge a young boy and his mother – who was armed only with an 18 month-old baby – were murdered by Federal snipers in a bungled and wholly inappropriate raid on a peace-

'Had we cruelly trampled the purveyors of documentary truth in our rush to the top?'

ful but highly paranoid white separatist family. The exact same snipers committed a similar atrocity a few months later at Waco, Texas.

So the Michigan Militia's concerns about the federal Government deserve to be treated with something other than arched eyebrows. The Weaver family at Ruby Ridge and David Koresh's parishioners at Waco were victims of liberal America – gunned down because they were peaceful; if odd, beliefs didn't fit the prevailing norm. Moore ought to be championing these under-dogs in the same way he champions victims of corporate globalisation, but he doesn't. They are, to him, just wackos. Indeed later in the film he includes a clip at two of the protesters who gathered at Ruby Ridge to support the Weaver family. Again he portrays them simply as gun nuts, with our contextualising the cause of their anger. As Randy Weaver once said to me, "The second amendment, the right to keep arms, isn't about hunting or target shooting; it's there to protect the people against a government that can become tyrannical against its own people."

But then Moore's film really takes off. Examining the Columbine High shootings, which he points out – occurred on the same day that "Clinton bombed some country the name of which nobody in the US could pronounce", he asks: what is it about America that produces so much gun violence? Last year 381 Germans were killed by guns. There were 165 gun deaths in Canada, 68 in the UK, 15,112 in Australia and 39 in Japan. In the US there were 11,127 "Are we homicidal by nature?" asks Moore. He does a vox pop on the streets of New York. "Canadians don't watch violent movies," suggests a passer by "There are no guns in Canada," suggests

ourselves in the film: McDowell's art March, top; Nick with Biggie and

Tupac, centre; the Maysles brothers with Mick Jagger in 1970's 'Gimme Shelter' above; Joe Ranson, below.

ppened at Columbine is a microcosm of what no throughout the world." It's as if this report ore sees himself as a bystander to these mat supposed to being at there very heart.

ore is a superb polemicist. Chomsky with a brilliant collagist of both kinds archive – heral polemics. After an opening mont hit documents America's obsession guns throughout its cultural and all history, he is seen walking into Twentieth bank "I want the account that I con't 'criminally defective'" 's OK if I'm normally men- hrks out but not criminally." Moore "Yes," says the bank ger "Don't you think it's a

another. It turns out, of course, that Canadians love violent movies and love guns. The answer is as flatulent as Moore. A problem with Moore's less successful work is his intellectual-irresponsive; his having all the answers there he seems honestly confused.

The turning point comes in an interview with pop star Marilyn Manson, who was absurdly scapegoated for inspiring the Columbine shootings because the boys were fans of his music. But the boys went bowling the morning they killed their classmates. Why

'Broomfield and I also arch our eyebrows at neo-Nazis, but I'm trying to stop doing this'

is no nobody blaming bowling? In the aftermath of Columbine children were expelled from schools across the US, one for pointing a chicken drumstick at a teacher "Yes, our children were indeed something to fear," says Moore.

"Ultimately *Bowling for Columbine* is a film about lear Americans are lear rankers and the media provides all the fear they want. Fear of Marilyn Manson, fear of out of control children, and – post 11 September – fear of everything, which sanctions US foreign policy outrages. In grappling with this concept, Moore transforms himself from a political commentator into a philosophical one. He suggests too is that the fostering of fear is a capitalist sleight of hand designed to deflect Americans' attention from the things they should be angry about – the things that make money for corporations.

"A country that's lost out of control with fear shoulders have all those guns and, ammo lying about." Moore con-

cludes, ending the film by going after the National Rifle Association president Charlton Heston. It's a disappointing climax, which attempts to ape his debut *Roger & Me*. but Roger Smith was a worthy target, and Heston is not. Sure, Heston was a little foolish in holding an NRA meeting down the street from Columbine High a few days after the massacre. But he's a lobbyist who's not responsible for the murders of little girls, and when Moore leaves a framed photograph of a murdered little girl on Heston's doorstep and wanders away with his head bowed, the pathos backfires. In demonising the NRA – he even accuses them of being the Ku Klux Klan by another name – Moore is himself guilty of inappropriate scaremongering. It's an unsatisfactory ending to an otherwise brilliant documentary. *'Bowling for Columbine'* is showing on screens 2–4 November at the BLFF, is released on 15 November and is reviewed on page 40. For details of the BLFF programme call 020 7928 3232 or visit www.lff.org

Arms and a man is investigating US attitudes to guns in 'Bowling for Columbine'; Moore, above;

draws statements from a representative of arms manufacturer Lockheed Martin, top, among others

Outsider: Audrey Tautou as Senay

Nick James on Stephen Frears' London immigrant drama

Subterranean homesick blues

Given how smoothly Stephen Frears seems to be able to hop from US-based movies like 'The Grifters' (1990) and 'High Fidelity' (2000) to British-scale features from 'My Beautiful Laundrette' (1985) to 'Liam' (2000), it ought not to be such a surprise to find him portraying the hidden workers of London. Yet 'Dirty Pretty Things' is startling in the correct climate because it's so unafraid of qualities which script-formula gurus advise against. It's a sunny urban thriller with an obvious special effects and a weighty political dimension. It stars a little-known male lead in Chiwetel Ejiofor (admittedly playing opposite such European names as Audrey 'Amélie' Tautou and Senji López) and is set in a downbeat milieu of the dispossessed, filmed with appropriate tension and bleakness by Chris Menges.

Ohwe (Ejiofor) is a Nigerian man, once a doctor but now ducking sleep to pull wrapt on two low-paid posts in London – nighttime at the seedy Baltic hotel and daytime minicab driver – with a further sideline in ministering to the STDs of his equally 'Vaseline' colleagues, the sleeps on a much

belonging to one of the Baltic's cleaners Senay (Tautou), a Turkish immigrant working illegally. When the attentions of the immigration inspectors force her out of her job, she's ripe for the victimisation. Ohwe feels responsible for her but seems powerless to help. Soon they are caught at the eve of a vicious whirlpool of deprivation.

The sense of a class of workers invisible to the citizens they serve but dependent on each other is deftly achieved. But you shouldn't get the impression this is a worthy film. It's an effective thriller made all the more urgent by the social concerns at its heart, and its horrific elements are as queasy and gripping as anything in 'The Grifters'. Some might find Tautou's inescapable cuteness a touch inappropriate at times, but she remains plausibly brittle and holds her own among this terrific cast. Most of all it seems like a film that could have come from the heyday of 1980s television-financed film-making that we seem to have lost sight of recently – except that the lesson it illustrates could not be more vital to the present. 'Dirty Pretty Things' is on 6 and 7 November

Doctor in the house: Chiwetel Ejiofor as Okwe

Reviews

Ryan Gilbey wonders if the politics of 'Changing Lanes' are shift stick or unthinkingly automatic

Street legal

Samuel L Jackson has not yet provided compelling evidence that he can play much besides funky hipsters (*Pulp Fiction, The Long Kiss Goodnight*) or righteous avengers (*A Time to Kill*), but the conscientious thriller *Changing Lanes* hints there are fresh ambiguities to be mined in the latter category. Here Jackson plays Doyle Gipson, whose abstinence from alcohol does little to temper his temper after a collision on New York's FDR Drive with hotshot attorney Gavin Banek (Ben Affleck). Doyle tries to act honourably. It is he who refuses Gavin's offer of a blank cheque, and later he will also make an attempt to return to Gavin the import ant legal

both instances Doyle's good nature goes unappreciated. In the first Gavin speeds off to court, leaving Doyle – who also has a court appointment, to stop his ex-wife and young sons from moving state – stranded on the freeway. In the second Doyle's altruism comes too late to prevent Gavin's visit to Mr Finch, a computer hacker who renders Doyle bankrupt with the touch of a button.

The screenplay, by Michael Tolkin (*The Player, Deep Cover*) and debutant Chap Taylor, sometimes seems poised to commit the ultimate heresy of making a main character in a Hollywood movie unsympathetic: in one scene Doyle clubs two casually racist strangers with

---

Theroux and I never hug, though Broomfield recently became a neo-hugger (see the final "you make a lovely stew" scene in *Biggie and Tupac*).

At his very best, in *Roger & Me* and now *Bowling for Columbine*, Moore is quite brilliant at creating – using archive material and savage comedy – a political panorama that startlingly interweaves the macro with the micro. One of the most powerful and convincing moments in *Bowling for Columbine* is his drawing of a parallel between the US selling weaponry to Eric Harris and Dylan Klebold, the teenagers who shot up Columbine High on

"I want the account where I get a free gun, says Moore. The bank manager calmly agrees"

20 April 1999, and the US providing training and finance to Osama Bin Laden during Russia's invasion of Afghanistan. He even manages to get a representative of arms manufacturer Lockheed Martin to say, "What ▶

广告印刷品

从很大程度上讲，这种类型的广告主要依靠强有力观念和良好艺术导向的组合。无论语言还是图片，都要让受众相信这件产品是为他们生产的。优秀的文案是很重要的，它必须让受众印象深刻。在此，作为设计师的你应该做的是通过字体设置和排版使信息清晰明白。

如果广告是用作推广或信息发布，你可以扮演更有影响力的角色。在文案撰写完成、需要突出的重点已经确认之后，你可以通过视觉元素反映文案的内容。

我们的文化造成了这样一种预期：广告应该是与众不同的。这给了你在设计中灵活运用字体并创新的机会。对图片和文字的创新组合往往会造就引人注目的广告设计。如果设计中仅靠文字而没有图片元素，色彩翻转、字体重叠、方向改变等都可以为解决问题提供参考。文案也可以用一种幽默的方式撰写，它让你通过自己的选择发掘其中的乐趣。前面用了大量的篇幅，这里就不对你应该在字体上进行的"疯狂"尝试进行赘述了，但广告设计确实是一个值得一试的设计种类。

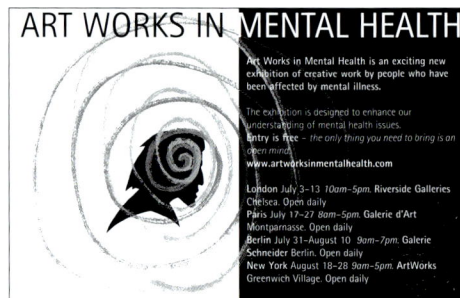

ART WORKS IN MENTAL HEALTH

www.artworksinmentalhealth.com

ART WORKS IN MENTAL HEALTH

Art Works in Mental Health is an exciting new exhibition of creative work by people who have been affected by mental illness.

The exhibition is designed to enhance our understanding of mental health issues. Entry is free – the only thing you need to bring is an open mind.
www.artworksinmentalhealth.com

London July 3–13 10am–5pm. Riverside Galleries Chelsea. Open daily
Paris July 17–27 8am–5pm. Galerie d'Art Montparnasse. Open daily
Berlin July 31–August 10 9am–7pm. Galerie Schneider Berlin. Open daily
New York August 18–28 9am–5pm. ArtWorks Greenwich Village. Open daily

- 设计一张名为"艺术对精神健康的影响"的广告。
- 它应占据一本娱乐导刊中展开的两个页面。
- 目标群体的年龄为 25—35 岁。
- 广告页面应采用风景格式的页面进行设计。
- 它必须包含至少两张你自己设计加工的图片及范例中列举的信息。
- 采用四色。
- 尝试图片、形状、字体和风格的不同组合。
- 下面的创意将帮助你着手设计。

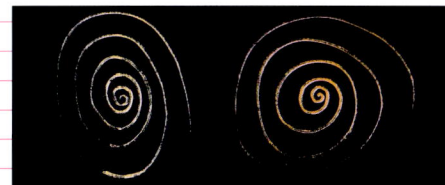

ART WORKS IN MENTAL HEALTH

Art Works in Mental Health is an exciting new exhibition of creative work by people who have been affected by mental illness.

The exhibition is designed to enhance our understanding of mental health issues.

Entry is free–*the only thing you need to bring is an open mind.*

www.artworksinmentalhealth.com

London July 3-13 *10am–5pm*. **Riverside Galleries** Chelsea.
　　　Open daily
Paris July 17-27 *8am–5pm*. **Galerie d'Art** Montparnasse.
　　　Open daily
Berlin July 31–August 10 *9am–7pm*. **Galerie Schneider** Berlin.
　　　Open daily
New York August 18-28 *9am–5pm*. **ArtWorks**, Greenwich Village.
　　　Open daily

广告印刷品
范例解析

　　构成广告的设计要素包括协调的基本形状、抢眼的"garbs"字样和商场的详细信息。这些要素组合起来使一系列设计立即跳入人们的眼帘，并且，更为重要的是，让人们过目不忘。

　　这个设计中最有趣的细节是将众所周知的词语换成与之读音相似但反映的是产品信息的另一个词语。例如，在箱包类的广告中，我们看到的是"stuff"，而不是"staff"。

　　排版调整的使用比较简单，但却体现了变化。对字母大小写、字体和标点符号的不同运用都增强了这一效果。字体的磅值也与产品的感觉相符——中等磅值的字体用在皮鞋和箱包的广告中，粗体字用在毛衣的广告中，加粗体则用在外套广告中。

1. 选择合适大小的标题字体可以起到平衡的作用，不同尺寸和风格产品间的直观联系通过线条得以表现。
2-3. 标明箱包设计者的标签也作为一个元素加入到设计中。在图片下方居中的文字表明，在这张整洁广告的主体部分，对称是可以实现的。
4. 这张图片有三维立体的效果，但仍然保持了简约性。黑色的毛衣置于顶部，颜色较浅的一件则紧靠其下。
5. 在标题中使用一个与其他文字颜色不同的"&"符号，意味着许多字母可以用较大的字体排列在同一行中，并且不至于拥挤，标题与当中的图片也相得益彰。

BOOTS TO SUIT.
In chocolate brown, camel and black leather.

High

Ankle

Shoe

garbs
clothing   accessories   jewelry

Ruby Isle 2205 North Calhoun Road, Brookfield
Monday-Friday 10:00-6:00, Saturday 10:00-5:00

262 780-0909

1

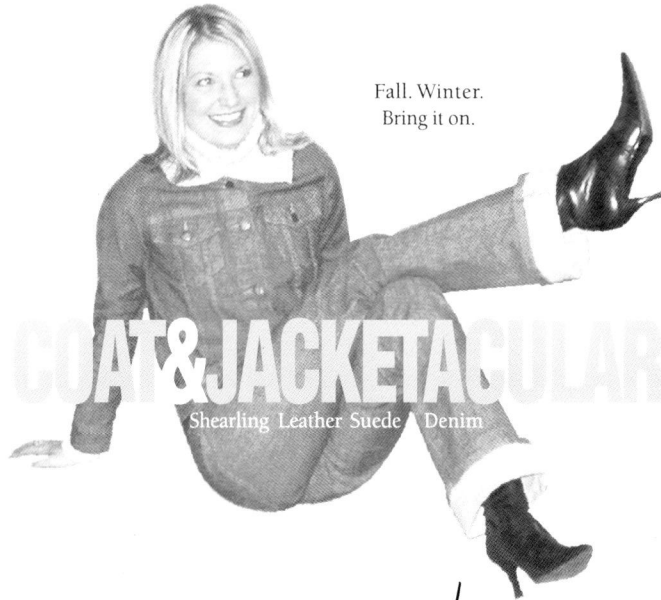

理查德·霍利斯在《平面设计》一书中说:《简史》中把海报描述为平面设计的基本形式,而它更常被描述为一个艺术形式,也是长期以来我们进行沟通交流的形式,是我们文化中不可分割的一部分。

海报

- 设计一张海报,告知一系列哲学讲座中的一场讲座。这些讲座是由亚里士多德学会组织的。
- 海报将在中学、学院、大学和图书馆中张贴,客户要求引用这一系列讲座中涉及到的哲学家的名人名言。因此引文应容易阅读,同时,你也要在设计中体现出引文的含义。
- 你可以操纵的元素是对字母的选择,以

从19世纪最早的海报(反映商会的早期斗争)到今天遍及欧美的平面设计的运用,海报的发展向我们展现了一部活生生的历史。

优秀的海报设计要发挥两个作用:首先,它要引人注目;其次它要严谨正式。好的海报设计均需要达到这两个标准,同时还要体现出对色彩的合理运用。海报是一块可供描绘的画布,为你提供了将字母当作图片进行描绘的机会。

在海报中需要谨记的是主要信息必须被周知。海报通常只是在人们经过时才投以匆匆的一瞥(试想路边的公示栏)。你可以用较小的字体设计一本书的封面,因为它的阅读距离只有6至8英寸,但海报设计中却千万不能采用小字体。同样,也不要在海报中包含过多元素,只需要通过形式和色彩的运用使重点鲜明突出。如果你研究一下多数效果明显的海报,你就会发现单一、强烈的要素总比多种不同的要素来得成功。受众喜欢清楚明白,当他们能够相对容易地了解一项信息时,他们通常会觉得比较舒服。

大多数情况下,海报的印数都不多。丝网印刷是较好的选择,因为采用这种方式印刷出的作品色彩鲜活,有光泽,体现得出金属质感,也可以在黑纸上印出浅淡的颜色。如果有上千张要印刷,可以用平版印刷的方式,虽然色彩会单调一些,但在印制四色作品时效果还是很不错的。

"Cogito ergo sum"

"Cogito ergo sum"

"Cogito ergo sum"

"Cogito ergo sum"

"Cogito ergo sum"

"Cogito ergo sum"

"Cogito ergo sum"

"Cogito ergo sum"

"Cogito ergo sum"
(I think, therefore I am)

RENÉ DESCARTES *1596–1650*

"Cogito ergo sum
(I think, therefore I am)

RENÉ DESCARTES *1596–1650*

及对它们的行列、组合、色彩和风格的设置。

- 从下列引文中选择一条：
  "任何人的知识都超不出他的经验"——约翰·洛克（1632—1704）
  "我思故我在"——笛卡尔（1596—1650）
  "世界就是一切，这就是现实"——路德维格·维特根斯坦（1889—1951）
  "人心是为了调和最为激烈的矛盾"——

休谟·戴维（1771—1776）
"因为它是圣地，才受到上帝的眷顾，还是因为上帝的眷顾，才使它成为了圣地？"——柏拉图（公元前427—347）

- 现在添加下列文字
  "亚里士多德学会"的系列讲座从[你选择的哲学家的名字]开始，在[时间]举行。欲知详情，请联系位于[大学名]的联系人。

- 将设计好的作品贴到墙上，并退后至合适的距离。设计的吸引力在这第一瞥中显得非常明显。询问自己是否有文字难以辨认，是否需要进行调整。

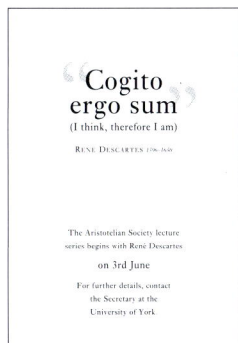

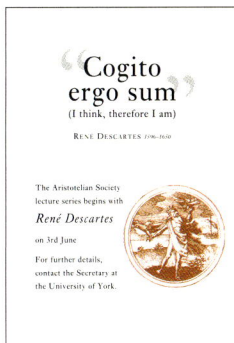

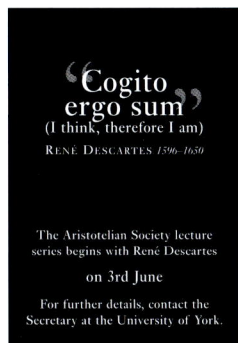

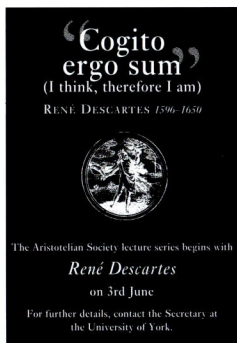

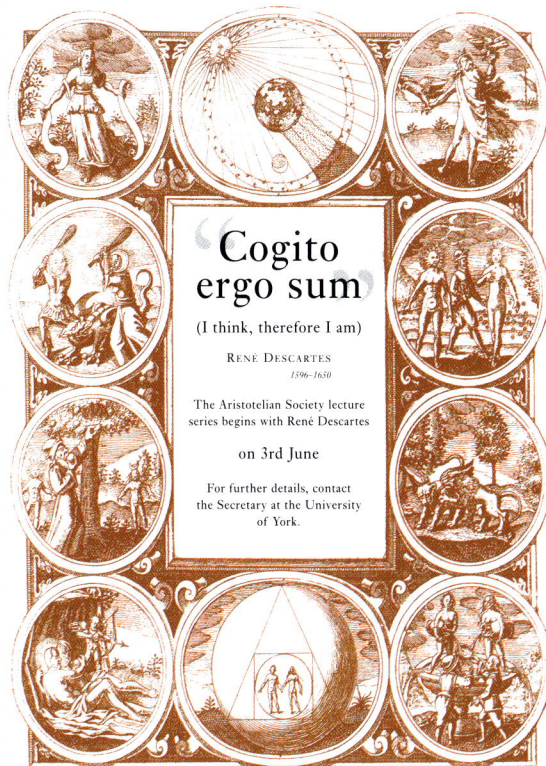

- 下面的创意将帮助你着手设计。

成功的海报并不企图通过文字和图片表达过多的信息，过于复杂的设计会引起混乱。相反，它们应该在构造、形式及更重要地在色彩运用上具有视觉吸引力。最后，记住它们应该具有告知性，最好在一瞥之间就能传递信息。

海报
范例解析

这张海报传递出的信息是清楚的，代表水滴的图片很好地切合了主题。选择的字体清晰明了，与图片也不冲突。

1. 简洁但却有效的组合。两者均传达了包含的信息并且互不干涉。
2. 采用无衬线字体的粗体，增加了文字的力量。由于字体较小，因此不会抢了图片的主导地位，反而为背景增加了一种现代感。水滴被放到了页面正中，突出了它们的重要性。
3-5. 在这里，设计师改变了背景的颜色。通过这几个范例，可以清楚地看到不同颜色对设计的不同影响。
6. 最后被采用的海报有着生机勃勃的色彩，设计中没有不必要的词语，信息通过图片表达出来。这项设计引人注目、简洁，并且让人难以忘记。
7. 模糊、显得有些聚焦不准的图片给设计增加了抽象的意味，居中的文字虽然小，但却一眼就能辨认出来。
8. 眼睛是这张海报中唯一未被涂写过的地方，正因如此，才吸引了受众的目光，文字也变得更具意象表达的效果，而不仅仅起告知作用了。
9-10. 两张海报均体现了设计取得成功的要诀：亮丽的色彩、简洁的图片和尽量少的文字。

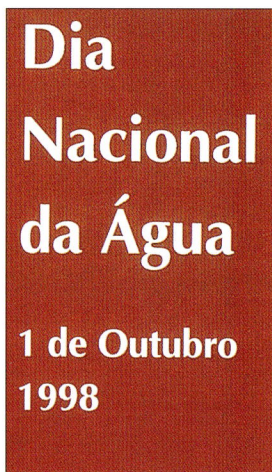
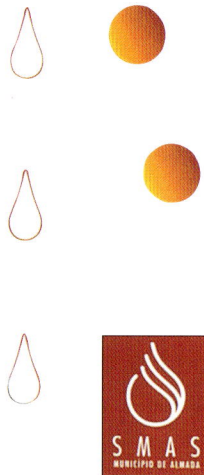

专业范例

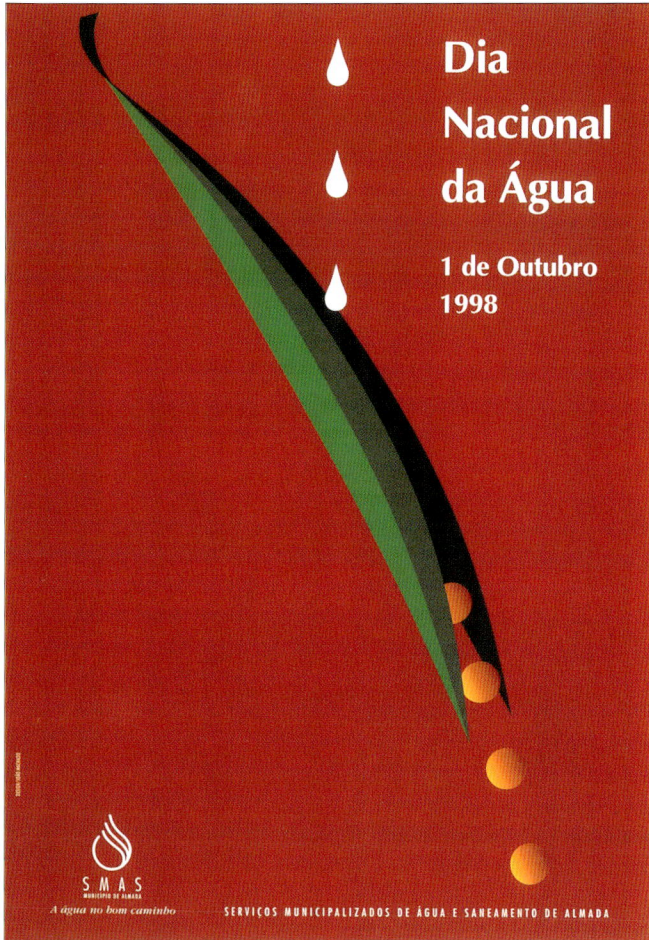

Dia Nacional da Água
1 de Outubro 1998

SMAS
MUNICÍPIO DE ALMADA
A água no bom caminho
SERVIÇOS MUNICIPALIZADOS DE ÁGUA E SANEAMENTO DE ALMADA

6

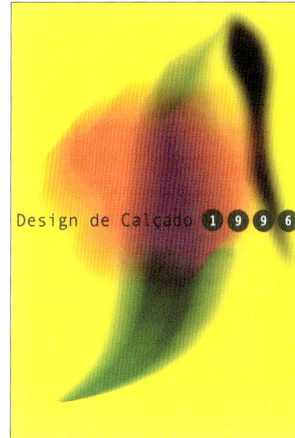

Design de Calçado 1996

7

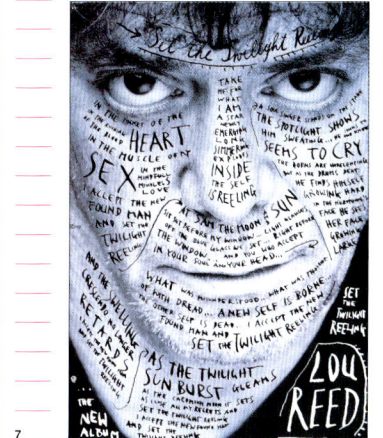

Set the Twilight Reeling
LOU REED

8

Cinanima 2000

9

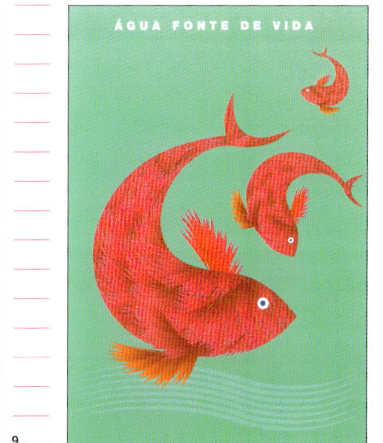

ÁGUA FONTE DE VIDA

10

包装与标签

　　某些包装仅仅需要具备装饰功能，而其他的包装，如医疗产品的包装，则需要传达重要的信息。这两种设计的基本原理明显是不相同的。在需要传递信息的包装设计中，清楚的字体是最重要的，你需要专注于尽量又快又清楚地表达信息。大小写混合的字体比全部大写的字体容易辨认，同样，左对齐的格式也比居中的格式易于辨认。避免使用右对齐、倾斜字体或装饰字体，或者将某些字黑白翻转。尽量不要断开单词。对于警告或其他重要的信息，你的目标是在色彩安排上使其与别的信息形成鲜明对比。你或许应该听取别人关于色彩的意见，特别是色盲及色弱人群。

　　装饰性的包装为你提供了更多创造性使用色彩、字体和图片的空间，你也可以对页面分配进行各种尝试。不对称的设计在这类设计中发挥了积极的作用。

　　在某些类型的包装设计中，你需要设想设计在不同表面上的效果。例如，你需要考虑如何让所有的文字和图片从包装的前面延伸到侧面。包装中的物品或许可以决定你所采用的颜色范围。通常产品的名字也会在选择颜色时给你以启发。

　　包装生产的方式也需要予以考虑，因为它们会为设计增加额外的要素。玻璃纸、彩色棉纸、去除背景的图片及其他辅助的印刷技术，如层叠、抛光，或浮凸图案等都是可行的选择。

　　设计完成后，在三维的实体模型上进行试验。这将会向你和客户展示整个设计是怎样的。

- 为"化学香精"和"赖德香薰蜡烛"设计包装表面和标签。
- 包装应朴素且现代，无需太精致，是为新千年的消费者设计的。
- 两件产品的价位均属中等。

- 包装的尺寸和形状由你自己决定，记住它必须与产品相符。
- 采用四色。
- 在作出最后的决定前，做几个实体模型。确保你选择的设计在包装和运输上都能

满足要求。
- 将包装设计成完全的三维格式。
- 下列创意将帮助你着手设计。

$$E = mc^2$$
$$a^2 = b^2 + c^2$$
$$x = 2y + z$$

技巧的使用为设计师提供了进行各种创意尝试的广大空间。

包装与标签
范例解析

图片CD可以通过图片、印刷品展示及高端训练等方式帮助摄影师了解各种不同的文化。因此这是一种区分信息的有效方式。包装通过采用聚丙烯的CD盒及半透明的纸张表现了信息的多层次性。

1. CD和包装中均使用了厚重的色彩，使得名字可以凸显出来。无衬线的字体也使得黑白翻转后的字体清晰可辨。
2. 整个包装具有现代、令人振奋的感觉，反映出了CD包含的内容。
3. 清晰简明的直线色条选用的是色谱中邻近的颜色，使购物袋的外观显得精致考究。
4. 为实用的物品设计实用的包装。
5. 这个现代的火柴盒其实是其中产品的形象刻画。
6. 繁杂而质朴的设计——形状和色彩的大集合。它反映出了产品的特征——声音的爆发。
7. 这是一个精心设计的CD标签和包装。白色背景中微小的图片营造了和谐的氛围，这种氛围还通过图片中的宁静气氛得到了增强。

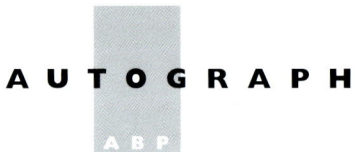

1

2

专业范例

3

4

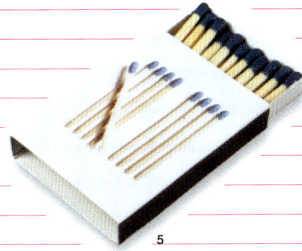
5

6

7

首页与链接

在作出任何决定时,想像站点在将来可能的发展前景。这时你或许应该为你的文件建立一套命名机制。随着网站的成长,如果不好好组织的话,要查找任何东西都很困难。例如,同一用途的文件使用相同的文件名是可取的。因此,如果一个文件名为 home.htm,那么插入其中的动画文件可以命名为 home.swf。

记住你的访问者并不只是沿着同一方向浏览站点,他们可能在不同问题间向前向后浏览。他们也可能是从其他站点的链接中进入你的站点,而并非通过主页。为此,你需要一套可以应用到任何页面中的简洁的系统。当点击时,机构标志总是带你返回首页,这是一个普遍接受的惯例。这个标志通常位于页面的左上角,站点的导航栏则位于标志下方或紧随标志之后,横贯整个页面顶部。

这样形式的导航栏意味着它总是可见的,你总可以在屏幕的左上角看到它。然而,如果某人的屏幕小于页面尺寸,任何位于底部和右面的信息都不能同时显示,他们需要通过滚动来查找。

链接将一个页面与另一个页面联系起来。同一页面内的链接的颜色通常在未被点击前比主色调浅,而在点击过后就变深了。惯常地,链接以下划线标明,虽然也可以不必如此。更为重要的是你不要使用任何会使内容看上去像链接的方式来强调某些文字,比如加上下划线或改变颜色等。通常采用磅值的差异来突出这些文字。

页面尺寸

因为不可能控制页面尺寸(页面的垂直和水平滚动是没有限制的),页面大小应根据屏幕尺寸进行调整,这通常通过使用无边框(因此看不见)的表格来实现。

字体

记住网页是生动的媒介,访问者会在其中跳来跳去,而不像阅读书籍那样逐行研究,任何文字通常都比印刷格式中来得简短概括,因为人们不喜欢老是上下滚动页面或者加速眼睛的疲劳。你选择的字体应与整体风格相符,并且应包含在常用字体中,如那些可以在 microsoft internet explorer 中显示的字体。

色彩

印刷页面通常采用 CMYK 印刷(见 62 页),计算机的显示器采用的是 RGB 格式的色彩,依据使用者显示器的分辨率,从 256 种色彩到百万种色彩都能显示出来。如果没有应用程序,你无法得知使用者显示器的分辨率,因此最好限制在 256 种色彩之内。在 256 种色彩中,只有 216 种在 Mac 机和家用机上都能显示。这 216 种色彩被称为网络安全色,也能在所有页面设备中显示。这 216 种色彩可以叫做 gif 格式。对照片及其他连续色彩的图片,则采用 JPEG 的格式以防止色彩混同。

- 你受命为一个名为"美丽花束"的网上花店设计网页。
- 首先在纸上勾勒出对三个页面的粗略构思：首页和两个链接页面。
- 用 photoshop 或其他类似软件设计页面，

- 你的工作区域是 10 × 10 英寸一页。
- 加入照片、图片和文字。
- 首页应显示至少五个链接，确保使用者能在你设计的页面中自由浏览。
- 选用的字体和色彩应反映行业的特征。

- 在左上角放置一个标志。
- 下面的创意将帮助你着手设计。

版面设计

你应避免使用列状文字，因为访问者不得不上下滚动页面进行阅读。但你却可以设置工具条，它通常链接到其他相关内容上，这也是利用空白空间的有效方式。

转换

设计完页面后，必须使用页面生成工具，如 Macromedia Firworks 或 Adobe Image Ready 等软件，将它们整合为 HTML 格式。尽量避免生成过大的文件：单个文件不要超过 50K。你的站点应囊括所有现行的浏览器和平台。如果需要使用者下载某个特殊的浏览器才能阅读你的内容，那么这个设计将是失败的。

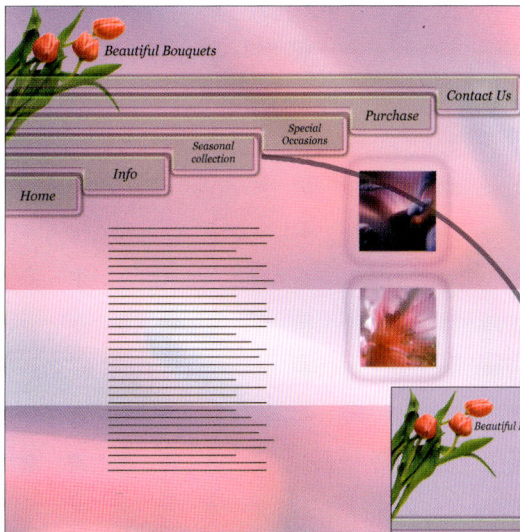

互联网站点只有在访问者能自如地往来于不同站点间时才是成功的。这也是为什么某些要素（如导航工具）在整个网站的页面中始终以相同的风格保持在相同位置的原因。

首页与链接
范例解析

此处的客户（一家图片代理商）需要一个清楚、时尚的网站，并且要能反映他们所从事行业的特征。网站中的文字极精简，所有的中心都在图片上。每张图片都经过了精心的剪切和处理，并根据不同尺寸调整了格式。页面是活动的，或独立或依靠鼠标点击，为网站增添了趣味性和创造性。

1. 客户需要一个能使人振奋的性感的主界面。
2. 为达到最佳效果，模糊的静态图片穿插到其他图片中，形成了一种速度感，而这仅靠动画效果是无法实现的。
3. 导航工具位于页面正中，以保持页面的平衡。但访问者可以将它拖动到屏幕的任何位置。半透明的条块衬着暗色调的背景，保证了菜单上条目的易读性。
4. 拖动时，菜单会像钟摆一样来回摆动，并且持续不断。
5. 这张页面反映了使用全屏显示图片的潜力，同时也体现了导航工具可以做到何等的精巧。菜单被重新放到了屏幕左端，访问者与其之间不再有互动，视线的焦点完全集中在摄影师的作品上。图片左端的两个小方块是"前进"和"后退"按钮，通过它们可以在同一系列的不同照片间进行切换。位于页面左下角的公文包状按钮是"退出"按钮。

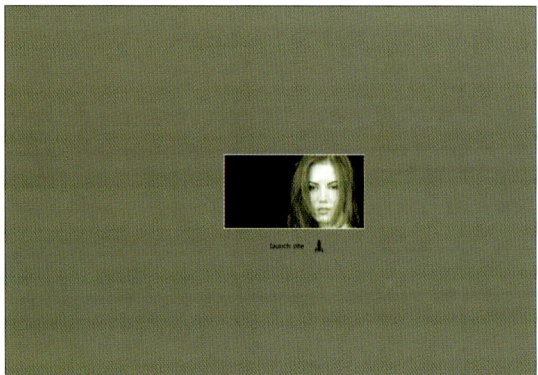

2

3

1

4

5

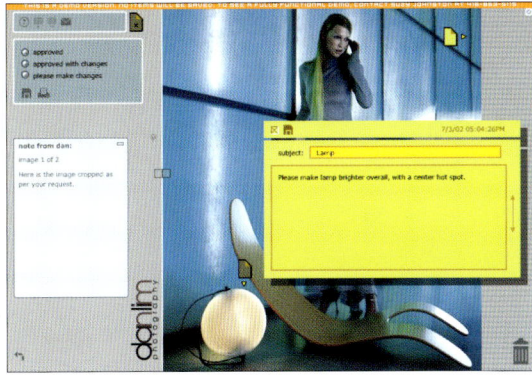

8

6

9

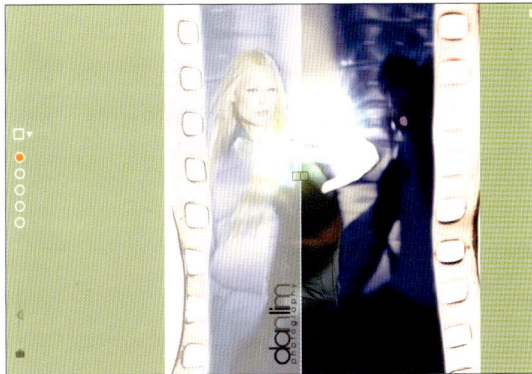

7

6. 点击"前进"和"后退"按钮，可以使现在的图片被从屏幕上"擦除"，显示另一张图片。

7. 这里显示了前进或后退时两张图片的切换情形。在这个例子中，半透明的"擦除"工具正在发挥作用。

8. 现有客户可以进入网站中受到密码保护的私人区域或数据库，并在那里发帖。

9. 在点击菜单中的"联络"选项后进入的界面中，访问者可以留下他们的联络信息，将格式设置成动画形式比HTML形式更能体现用户友好界面，因为在提交用户信息后，用户仍能停留在该界面中进行进一步的确认。

首页与链接
专业范例

1. 为这个专业卫生洁具销售商制作的主页非常清楚。由于图片的居中,设计更具现代感,标志出现在每一页的左上角,链接总位于标志下方,使得从任意位置进入每个页面都很容易。

2-3. 每张页面都遵照相同的基本格式,突出一条简单的信息。这是一个经过深思熟虑后建立的精致的网站。

4-7. 这个为促进兽医研究而设计的网站需要插入动画效果。这是一个朴实、色彩丰富、生动活泼的网站,圆形构成了网站设计的基础:从垂直排列的圆形链接按钮到圆形的字块,及当你将鼠标移至红色和白色的圆圈时,页面会相应上下滚动等。虽然有时常变换的圆形图片和复杂的构架,在这个网站中进行浏览还是比较容易的。

8-10. 这是个为图片库建立的简洁的网站,图片库收集的是花草、树木的照片资料。背景色彩是柔和的,网站也具有一种宁静自然的感觉,很好地反映了自然摄影的特色。

4

5

The Use of Plasma Cardiac Troponin I (cTnI),
Cardiac Troponin T (cTnT) and Atrial
Natriuretic Peptide (ANP) as Biochemical
Markers of Cardiac Injury Associated with
Doxorubicin Chemotherapy

6

7

flowerphotos

Welcome…

to a fascinating photo library.

We specialise in creative shots of flowers, plants
and trees; and more recently fruit and vegetables.

To review your lightbox, click here

Tel: 020 7684 5668 - 71 Leonard Street, London EC2A 4QU

8

9

10

设计师的角色在过去的几年中发生了巨大的变化，从很大程度上讲，计算机已替代了许多精细复杂的手工劳动。计算机打印出的图片拥有极高的质量，并且由此在人们心目中形成了一个显著的标准，人们也习惯于看到这样的作品。

向客户进行的介绍

当然，传统的技艺仍然在介绍时发挥着重要的作用。为背景剪裁边框、连接页面、处理构成设计的各种要素等都是做好清楚明白的介绍不可或缺的一部分。从心理学上讲，高质量的介绍是非常重要的，因为正是在那时，在那第一眼，印象的好坏就已经形成了。

在这之前需要提出的一个最为重要的问题是客户希望听到何种水平的介绍。如果他或她对设计过程非常清楚，那么一个使用黑白打印稿的"粗糙"介绍就足够了。对另一些客户，只有彩色打印稿才会有效。不管你选择哪种方式，你都需要同客户商讨时间和经费问题。他或她希望看到多少次介绍？在你对设计成本进行整体评估时要考虑到这一点。估算一下有多少东西要向客户介绍。显然，如果只是要求设计一系列带有标志的信笺，问题就简单了。如果是长达84页的小册子，你有必要向客户展示它的每一页吗？

当你准备介绍你的设计时，回顾一下它是否达到了最初与客户沟通时达成的目标。作为设计师，你必须培养良好的沟通技巧和能力，因为这是你推销作品的手段。你必须学会积极地倾听和回答。当然失误是在所难免的，但重要的是你处理失误和批评的方式。如果客户能看到你正努力解决问题，友善地解决任何问题都将非常容易。

最后，当你准备介绍你的设计时，要保证每个部分的项目都清楚地予以标注，并且按照正确的顺序排列。在介绍前花点时间远远比介绍后花两倍或三倍的时间去弥补过失要值得。

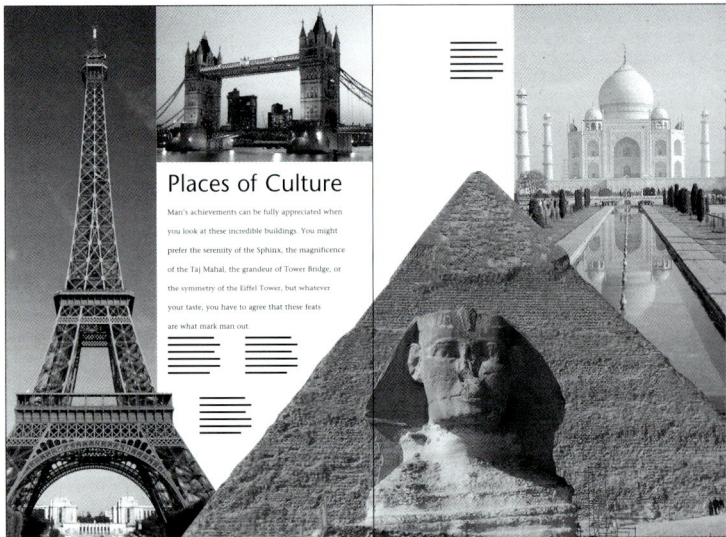

Places of Culture

Man's achievements can be fully appreciated when you look at these incredible buildings. You might prefer the serenity of the Sphinx, the magnificence of the Taj Mahal, the grandeur of Tower Bridge, or the symmetry of the Eiffel Tower, but whatever your taste, you have to agree that these feats are what mark man out.

- 在这个练习中，请你选择一个在杂志或小册子的设计练习中完成的项目，分三步向一个同事进行介绍。
- 首先，回到设计最基础的状态，在一张纸上勾勒出最初的创意。
- 然后，组织一次"粗糙"的介绍。这是一张没有图片和文字的最基础的页面结构图。
- 这次要实现的是设计完全完成后的介绍，所有的元素已经组合在一起，色彩也全部加上。
- 向你的朋友或同事介绍你的设计，并说明你实现最后的完整作品所要经历的几个步骤。
- 下面的创意将帮助你开始练习。

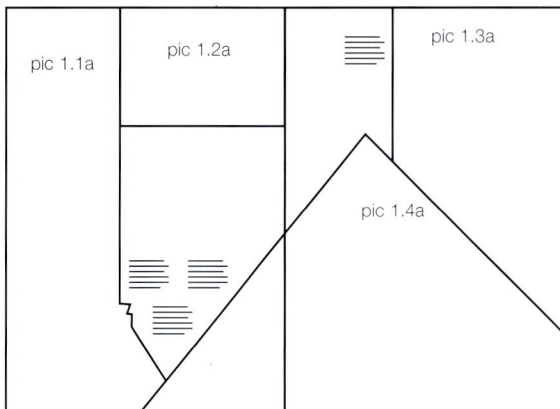

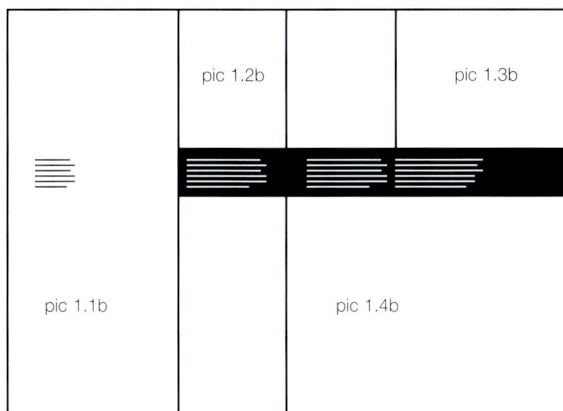

pic 1.1a
pic 1.2a
pic 1.3a
pic 1.4a

pic 1.2b
pic 1.3b
pic 1.1b
pic 1.4b

Places of Culture

Man's achievement can be fully appreciated when you look at these incredible buildings. You might prefer the serenity of the Sphinx, the magnificence of the Taj Mahal, the grandeur of Tower Bridge, or the symmetry of the Eiffel Tower, but whatever your taste, you have to agree that these feats are what mark man out.

Index

Index and Credits

t=top, b=bottom, c=center, l=left, r=right

P.1 bl The McCulley Group, br Studio International; P.2 t Chronicle Books, b The McCulley Group; P.3 l R2 (Ramalho & Rebelo Design); P.6 tl Chronicle Books, cl The McCulley Group, bl Becker Design; P.7 tc Einar Gylfason, t and cr The McCulley Group, bl Chronicle Books, br The Riordon Design Group Inc; P.10 cl Chronicle Books, bc Studio International; P.11 lc and c The McCulley Group; P.15 tl Terrapin Graphics, tr Yee Design; P.17 t Falco Hannemann, l Terrapin Graphics, cr Noah Scalin/ALR Design, br Eugenie Dodd Tyographics; P.21 tl Dotzero Design; tr Gee + Chung Design, bl Hornall Anderson, br Philip Fass; P.25 tl Chronicle Books, tr Fabrice Praeger; P.29 tl The Riordon Design Group Inc, tr Kurt Dornig, cl R2 (Ramalho & Rebelo Design), bl Chronicle Books, br Falco Hannemann; P.33 t, c, br Chronicle Books; P.36 tr Erich Brechbuehl; P.37 tl Kurt Dornig, tr The Riordon Design Group Inc, bl and r LOWERCASE Inc.; P.41 l Gee + Chung Design, r João Machado; P.45 tl and r Philip Fass; P.53 l Gee + Chung Design, r Grundy & Northedge; P. 57 t and bl Chronicle Books, br Studio International ; P.61 tl, bl and r R2 (Ramalho & Rebelo Design), tr Fabrice Praeger; P.65 tr Philip Fass, bl and c The McCulley Group, br Yee Design; P.71 r Gee + Chung Design; P.73 tr Noah Scalin, ALRdesign,

tl LOWERCASE Inc, br Studio International, bl The McCulley Group; P.77 tr and br Office Pavilion, tl, cl, tc Chronicle Books; P.81 tl and tr, bl LOWERCASE Inc, br Noah Scalin, ALRdesign; P.85 tl João Machado, tr Yee Design, br The Riordon Design Group Inc; P.86 t and cl Chronicle Books, cr The Riordon Design Group Inc; bc The McCulley Group; P.87 t and b Chronicle Books; P.92 tr and br R2 (Ramalho & Rebelo Design); P.93 l R2 (Ramalho & Rebelo Design), ct Eugenie Dodd Typographics, tr Hornall Anderson, c Becker Design, br Eugenie Dodd Typographics; P.96 Einar Gylfason; P.97 bl Einar Gylfason, tr Grundy & Northedge, br Supplied by author; P.100 Gee + Chung Design; P.101 tl LOWERCASE Inc, tc Lippincott, tr, b Eugenie Dodd Typographics; P.104–105 British Film Institute; P.108–109 Becker Design; P.112 João Machado; P.113 l, tc, bc, br João Machado, tr Sagmeister Inc; P.117 tl and tr Hornall Anderson Designworks,c LOWERCASE Inc., bl and br Sagmeister Inc; P.120 All The Riordon Design Group Inc; P.121 The Riordon Design Group Inc; P.122 picktondesign ; P.123 t and bl Mirko Ilic Corporation, t and br Flowers & Foliage.

While every effort has been made to credit contributors, we would like to apologize in advance if there are any omissions or errors. All other photographs and illustrations are the copyright of Quarto plc.